ONLY THE DEAD

ONLY THE DEAD

THE PERSISTENCE OF WAR IN THE MODERN AGE

BEAR F. BRAUMOELLER
The Ohio State University

OXFORD
UNIVERSITY PRESS

Oxford University Press is a department of the University of Oxford. It furthers the University's objective of excellence in research, scholarship, and education by publishing worldwide. Oxford is a registered trade mark of Oxford University Press in the UK and in certain other countries.

Published in the United States of America by Oxford University Press
198 Madison Avenue, New York, NY 10016, United States of America.

First issued as an Oxford University Press paperback 2021

An Online Appendix is available at https://dataverse.harvard.edu/

Library of Congress Cataloging-in-Publication Data
Names: Braumoeller, Bear F., author.
Title: Only the dead : the persistence of war in the modern age / Bear F. Braumoeller.
Description: New York, NY : Oxford University Press, 2019. | Includes bibliographical references.
Identifiers: LCCN 2019003397 | ISBN 9780190849535 (hardback) |
ISBN 9780197624272 (paperback)|
Subjects: LCSH: War–Causes. | War–Prevention. | International organization.
BISAC: POLITICAL SCIENCE / Civics & Citizenship. | POLITICAL SCIENCE /
International Relations / Diplomacy.
Classification: LCC JZ6385 .B73 2019 | DDC 303.6/6–dc23
LC record available at https://lccn.loc.gov/2019003397

1 3 5 7 9 8 6 4 2

Paperback printed by LSC Communications, United States of America

"Only the dead have seen the end of war."

–Santayana

CONTENTS

PREFACE

"So, are we seeing less war in the world today than we used to?"

It was spring of 2008 and I was standing in front of a hundred undergraduates, giving the first lecture of a new class on the causes of war. Such courses generally rely on historical case studies for evidence, and this one was no exception. At the same time, professional research in international relations often involves data analysis, so I wanted to give my students a few examples of how to use data to explore interesting questions.

I talked about a few different ways that one might think about the question. Are fewer countries fighting wars with one another than had fought them previously? Are the wars that they do fight deadlier? Are countries less likely to respond to provocations than they had been in the past? Have the causes of war disappeared? These are all valid ways to think about a decline of war, and they're all interesting questions. But we can't pursue all of them simultaneously, and it's important to narrow down which one you want to answer before you set about answering it.

In the end, I settled on a single question—how often do countries initiate conflict with one another? The deadliness of war is also an interesting question, of course. But the question of how often countries resort to violence in their dealings with one another struck me as being most relevant to the issue of whether war was in decline. There's a fundamental difference between disagreements that turn violent and those that do not, and I wanted to see whether the former were becoming less common.

Toward that end, I introduced my students to the Militarized Interstate Dispute dataset, political science's most venerable database of conflicts, going back to 1816. Militarized interstate disputes (or MIDs) are threats, displays, or uses of force—everything from saber rattling all the way up to all-out war. As a first cut, I simply added up the number of militarized interstate dispute initiations that occurred in each year and plotted them over time. I've reproduced that graph here directly from the lecture slide, as Figure 0.1. According to that graph, the world looked like an increasingly dangerous place.

"Looks pretty terrible, doesn't it?" I said. "I mean, just eyeballing it, it looks like the number of dispute initiations has increased at least tenfold over the last couple of centuries. Maybe more." Most of the students nodded in agreement.

"But let's think about this for a moment," I said. "What's wrong with this picture? What does it leave out?" I waited probably ten or fifteen seconds—it felt like an eternity, but if you start answering your own questions in this business, the students quickly realize that they don't have to. So I sweated it out. Finally, a young woman raised her hand and said, "The number of countries in the world has been increasing at the same time."

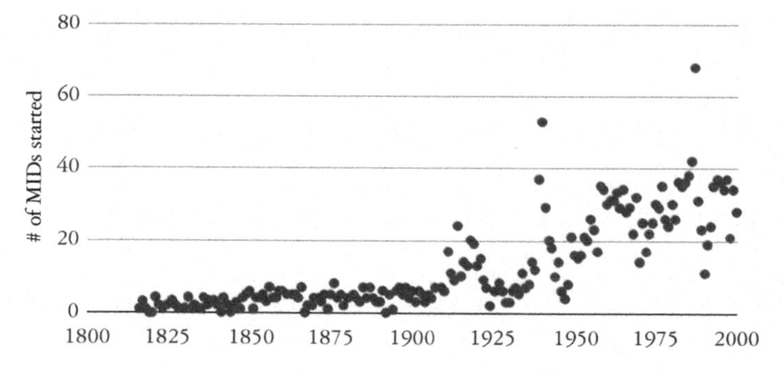

FIGURE 0.1 Militarized interstate disputes over time.

"Bingo," I replied. "If we're interested in measuring how often countries clash, we have to worry about how many countries there are in the world. If that number is increasing, we'd expect to see the number of conflicts increase too, even if states are no more warlike than they used to be."

I put up a graph that showed the increase in the number of countries in the world over time, and then I walked them through how to calculate the number of *pairs* of countries there are in the world—a number that grows much more quickly than most people think it does.[1]

To drive the point home, I said, "Let's imagine that every pair of countries has a 1% chance of fighting in a given year. If there are ten countries, or 45 pairs of countries, you'd expect a clash every couple of years. If there are 50 countries, or 1,225 pairs of countries, you'd expect about twelve such clashes every year. So when the number of countries increases, you should see a *lot* more conflict, even if the probability of fighting for a given pair of countries hasn't changed. Does that make sense?" Lots of nods.

On the next slide, which I've reproduced here as Figure 0.2, I divided the number of militarized interstate disputes in a given year by the number of pairs of countries that existed in that year.

"Now what do we see?" One student pointed to what looked like a decline in the first part of the Cold War and in the years that followed it. Another saw what might have been a slight increase in the latter part of the Cold War, though it could easily have been no change at all. A third

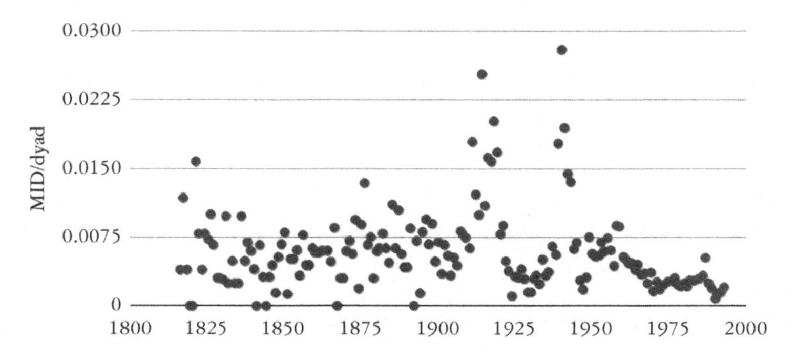

FIGURE 0.2 Militarized interstate disputes per pair of countries.

pointed out that the picture was surprisingly noisy prior to the First World War. All valid observations.

"So are we finished teasing out trends in warfare?" I asked. "Is this our final answer?"

I suspect no undergraduate has ever been tricked by a question like this. Mine certainly weren't: they could smell the booby trap a mile away. They scrutinized the graph and tossed out a few halfhearted answers until finally a young man said, "A lot of those countries can't reach each other?"

"Right," I replied, smiling. (I love my job, and *a-ha!* moments like this one are among my favorite parts of it.) "There's a pretty good explanation for why Bolivia and Botswana don't go to war with one another—they couldn't reach one another if they tried." Admittedly, it is true that the history of warfare has witnessed some surprising opponents. Colombia and Ethiopia, for example, each lost more than 100 soldiers in combat in the Korean War. But if you were going to bet on which countries would go to war with which others, more proximate countries are a much better bet than more distant ones, all else being equal. And as the international system gets bigger, the countries that make it up get farther and farther apart.

"So how do we fix *this* problem?"

The students offered a couple of ideas, and in the end someone hit on the solution that was waiting on the next slide: divide the number of militarized disputes by the number of pairs of countries that can unambiguously reach one another. The third graph, which showed the number of militarized disputes per pair of contiguous countries (Figure 0.3), was a little harder to interpret: it looked as though conflict might be declining after World War II, but it also didn't look as though the Cold War had been much less warlike than the nineteenth century.

"Is there a way to improve on this?" I asked. After a couple of false starts, one of the students pointed out that some countries, like the United States, are fully capable of starting wars with faraway countries, so those pairs should be counted as well. I responded by asking about regional powers—countries that can project power within their region but not globally. The students agreed that those pairs should be counted, too.

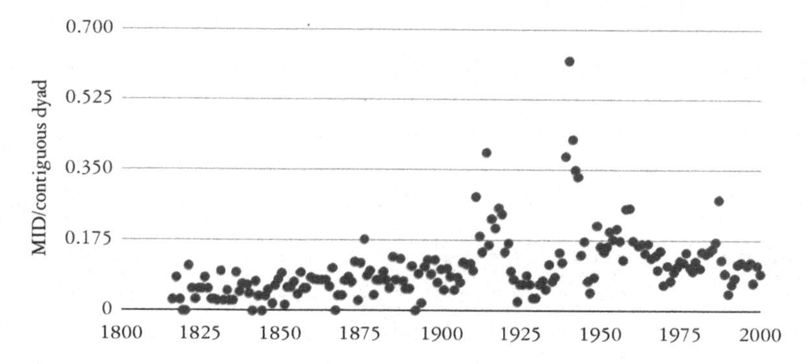

FIGURE 0.3 Militarized interstate disputes per pair of contiguous countries.

"You've got exactly the right idea," I said. "But this is where we stop for today. Figuring out which pairs of countries should be included and which shouldn't is actually a pretty hard question, and the more careful you want to be about it, the harder it gets." (I didn't add that it had been more than I wanted to tackle at 2 a.m. the night before, when I was writing the lecture. But that's true, too.)

In the end I said, "If this answer leaves you frustrated and dissatisfied, and you think you'd like to improve on it, you should really consider going to graduate school in political science. This is exactly the sort of thing we work on." After that shameless plug, I went on to show them how they could use data to explore a few more interesting questions about international conflict. Then I wrapped up the lecture, walked back to my office, and more or less forgot about the question of whether or not war was in decline.

I used those slides in subsequent versions of the same class, but I never refined the analysis further. I wanted to show my students the process of coming up with an interesting research question one step at a time, but I also wanted to leave them with that nagging itch that drives people to pursue a question until they've found a satisfying answer.

Also, I have to confess that the question of whether interstate conflict was declining just didn't interest me very much. I suspected it would take awhile to nail down the answer in a way that I'd find satisfactory, and as an untenured professor wrestling with an unfinished first book

I had other priorities. So for a while, I left it at that—a few late-night scatterplots that were more suggestive than conclusive.

The Decline-of-War Thesis

I knew, of course, that the argument that war is in decline had been made before. The person most often associated with the argument was Sir Norman Angell, whose 1910 book *The Great Illusion* put forward the thesis that war was futile, if not actually impossible. (The distinction was mostly lost on subsequent generations.) The surprisingly long stretch of time since the last Great Power war prompted similar works by Victor Hugo Duras (*Universal Peace*, 1908) and W. L. Grane (*The Passing of War*, 1912).

I also knew that nearly twenty years earlier my colleague John Mueller had written a book entitled *Retreat From Doomsday*, arguing that major war among industrialized countries had been on the decline for decades. He argued that war, like slavery and dueling before it, was becoming a thing of the past, as more and more people found it simply unthinkable. I also knew that he had extended this argument to cover the rest of the world in his 2007 book, *The Remnants of War*. John is a brilliant and iconoclastic scholar with a strong contrarian streak: A colleague of his once told me, "If you're ever looking for a great argument for something, just walk up to John and tell him you'd been thinking exactly the opposite thing. He'll come back at you with three great arguments in favor of the original thing." I had read John's books, and although I was skeptical of the bottom line, they were very smart. Anatol Rapaport's book *Peace: An Idea Whose Time Has Come* had been published in the immediate aftermath of the Cold War. It had not argued that world peace was at hand, but rather that human beings had it within their power to discredit the institution of warfare. Raimo Väyrynen's edited volume *The Waning of Major War*, which came out in 2006, had offered a useful variety of perspectives on John's thesis. None of them approached the question in the way that I would have, exactly. But I just didn't feel the urge to tackle the question myself.

I had no inkling that the decline-of-war thesis would go any farther than it already had. I didn't realize that another book on the same theme, Joshua Goldstein's *Winning the War on War*, would appear three years later. And I never would have guessed that, in the same year, Steven Pinker would publish *The Better Angels of Our Nature,* a book in which he argued not just that all forms of violence were in decline but that they had been for centuries. *Centuries!* Or that he would follow it up seven years later with *Enlightenment Now,* a more detailed argument that the Enlightenment ideals of science, reason, and progress were responsible for global increases in peace and prosperity and that, accordingly, what the world needs is more of the same.

Unlike Goldstein's and Mueller's books, which were best known among academics, Pinker's *Better Angels* was an instant bestseller. As a result, over the course of just a few years the decline-of-war thesis became a vastly more ambitious argument and reached a worldwide audience. That audience, for the most part, would find the decline-of-war thesis very compelling. In fact, it would achieve a nearly taken-for-granted status in the mainstream press. A typical review (Singer, 2011) concludes, "Pinker convincingly demonstrates that there has been a dramatic decline in violence, and he is persuasive about the causes of that decline." Many international relations scholars also found the argument compelling: Columbia University's Robert Jervis concludes that "he makes a case that will be hard to refute. The trends are not subtle—many of the changes involve an order of magnitude or more" (Jervis, 2011). While other scholars have been very skeptical of specific aspects of the decline-of-war argument,[2] the decline-of-war thesis has persisted and now forms one of the core ideas espoused by a group of pundits and commentators that Oliver Burkeman (2017) calls the "New Optimists."

The fact that a question at the heart of my academic specialty had suddenly become dinner-table conversation, combined with the fact that the growing conventional wisdom about the answer struck me as pretty implausible, finally piqued my interest. I started thumbing through the books and articles that made up the more recent decline-of-war literature. After a little while, I found myself jotting notes in the margins. Before long, the notes had gotten more frequent and more emphatic. I underlined questionable arguments and debatable conclusions; I circled examples of evidence that merited further scrutiny; I started to dig up some data and some papers on the subject.

Before I knew it, the itch that I'd tried to inflict on my students had started to nag at *me*. As the critical praise for Pinker's book and the decline-of-war thesis piled up, I became more and more convinced that a response was necessary.

At first, I thought I'd be able to satisfy the itch by putting together an "Author Meets Critics" panel at a professional conference on the subject of the end-of-war thesis, listing my complaints, listening to responses, and moving on. I contacted Professors Pinker and Mueller, who agreed to join the panel; I also enlisted Professors Bruce Russett, John Mearsheimer, William R. Thompson, and Andrew Mack. Unfortunately, our plans to meet at the American Political Science Association's (APSA) annual conference in late August of 2012 were upended when Hurricane Isaac forced the cancellation of the entire conference.

Next, I set out to do what academics do when we're annoyed about something: write a paper. Here, I faced something of a dilemma. I'm not unaccustomed to giving or receiving criticism—it's an essential part of the job. But criticisms of the number and gravity that I had started to amass *were* new to me, and I wasn't at all sure how directly I should address them. As I was weighing the possibilities, I came across a passage written by the late essayist and critic Christopher Hitchens (2005):

> I have been reviewing books on history and politics all my life, making notes in the margin when I come across a wrong date, or any other factual blunder, or a missing point in the evidence. No book is ever free from this. But if all the mistakes and omissions occur in such a way as to be consistent, to support or attack only one position, then you give the author a lousy review.

As rules of thumb go, this one has a lot to recommend it. It recognizes the fact that people make mistakes, it doesn't require attribution of intent, and it establishes a simple and reasonable criterion for a negative review. By that criterion, *Better Angels* certainly qualified for such a review. Nearly all of my marginal scribbles pointed to problematic claims that favored the central argument. Not one pointed in the other direction.

When I began to explore the data myself, the initial results confirmed my suspicions that the decline-of-war thesis just didn't square very well with the historical record. I delivered three very preliminary (and, in retrospect, very bad) versions of that paper at the following year's International Studies Association, APSA, and Peace Science Society International conferences and was surprised and gratified when the APSA paper was written up in *Popular Science* and *National Geographic.* Word got out, and I was invited to give talks at Yale, UCLA, Vanderbilt, the University of Chicago, the University of Pittsburgh, and the University of Iowa. The people who came to all of those talks helped to make the project much better than it had been, for which I'm very grateful.

In the early versions of the paper, I thought this would be a very quick project. I study war for a living, using statistics, so most of what I do involves estimating the impact of one thing on another. Using historical, observational data, that can be really challenging. This, by contrast, seemed almost trivial: All that was necessary was to measure a single variable and ascertain whether its value was increasing or decreasing over time. Maybe two variables, tops. I couldn't imagine squeezing more than a short paper out of it.

Slowly but surely, I was disabused of that notion. The process by which the scope of the project inexorably grew reminds me of a passage in Clifford Stoll's book *The Cuckoo's Egg.* Stoll recounts the experience of a graduate exam in astronomy in which one professor asked him a simple question: "Why is the sky blue?" After he replied that the atmosphere scattered sunlight in such a way that blue light was what we ended up seeing, the professor asked, "Could you be more specific?" After the next answer, which involved air molecules and the dual wave-particle nature of light, the professor again asked, "Could you be more specific?" As Stoll (1989, 339) tells it,

> An hour later, I'm sweating hard. His simple question—a five-year-old's question—has drawn together oscillator theory, electricity and magnetism, thermodynamics, even quantum mechanics. Even in my miserable writhing, I admired the guy.

So it was with this project. At first I thought I could answer the question of whether or not war was in decline by simply calculating the rate of conflict initiation and the intensity of war, plotting them over time, and running some simple statistical test to ascertain whether any changes had taken place. I soon realized that the data were too skewed for standard tests to be of much use. Those tests are designed, ideally, for nice, neat, normally distributed data, and what I had was anything but that. Reading Nassim Nicholas Taleb's book *The Black Swan* solidified my conclusion that I'd need to learn more about statistics before I could get reliable answers from the data. That process turned out to be surprisingly time-consuming but also really rewarding. In the end, I found everything I needed except a test for a difference in power-law coefficients, but such a test turned out to be straightforward to construct. (The details, along with much else that would only be of interest to specialists, are in the Appendix at the end of the book.) Robert Axelrod read an early and very rough draft of the analysis at this point and offered some very useful comments, for which I'm grateful.

Along the way, Dave McBride of Oxford University Press dropped me a line to see whether we could chat at a conference. Dave had no inkling that I was working on this project, but when I described this one his eyes lit up. He'd just finished *The Better Angels of Our Nature* and saw real potential for the skeptical rejoinder that I outlined. His last words as we parted—"I cannot urge you strongly enough to write that book"—are not the sorts of words that an author easily forgets.

As I was wrapping up the statistical analysis, I received an invitation to take up a fellowship at the Norwegian Nobel Institute in the summer of 2016 and present my research at a seminar on the causes of peace. The Institute turned out to have, as you might expect, a fantastic library on anything related to peace, and I spent a month blissfully working through revisions and extensions while my wife explored the city of Oslo with our infant daughter and enjoyed the hospitality of the scholars' flat at the Institute itself. The conference consisted of some genuine legends from the fields of political science, economics, history, and sociology, among other disciplines. Unfortunately, Professor Pinker wasn't able to attend, but I ended up on the opening panel with, of all people, Nassim Taleb, who was also in the process of publishing work critical of the decline-of-war thesis (Cirillo and Taleb, 2016*a*). Despite his fierce demeanor on social media, Taleb turned out to be incredibly gracious and generous with his time. We spent hours in the hallways of the conference hotel, comparing notes on the estimators that we'd derived.

When our wives stopped by and demanded that we stop working and come out to have some fun, we both replied, "This *is* fun!" And it was.

The rest of the seminar was one of the most stimulating experiences of my professional life. Quite a few people provided exceptionally helpful comments, either during the panel or afterward. Paul Diehl, Kristian Gleditsch, Olav Njølstad, Niall Ferguson, Geir Lundestad, Asle Toje, Ayşe Zarakol, and Bruce Russett stand out in my mind as having had especially useful feedback. Fred Logevall, who sat next to me during the seminar, offered some invaluable nuggets of wisdom and turned out to be a remarkably nice guy as well. He deserves credit for both.

Once the manuscript was finished, two anonymous reviewers, Rick Herrmann, Andy Dehus, Andy Moravcsik, and Kristen Schmidt were all kind enough to read part or all of the manuscript and offer useful comments. When he heard that it was nearing completion, Thom Winningham wrote out of the blue and asked whether I needed help with any of the computer-related parts; he was kind enough to re-render Figure 2.2 as a vector drawing, which improved it greatly.

Finally, Ms. Schmidt, who goes by Mrs. Braumoeller socially, put up with a lot of late nights and distracted answers during the writing of the book and did so with unfailing good grace. I'm not sure I would have taken on a project like this without her encouragement and support. I've also benefited greatly from her guidance on how to write like a normal person rather than a social scientist. The book is dedicated to her, for all these reasons and more.

In the initial presentations of this research some people concluded that, because I argued that war was not in fact disappearing, I don't *want* war to disappear—or, more charitably, that my own theories about how the world work implied that it would always be a dangerous place, and I had somehow selected only findings that would support that conclusion. Nothing could be further from the truth. I'm not a fan of war. The desire to help make it scarcer is a big part of the reason I got into this job in the first place. I don't have a strong commitment to any theoretical perspective that would preclude a decline in war. In short, I've got no dog in this fight.

I am, however, very committed to the notion that we should expose our ideas to relevant data in order to rule out some claims, support others, and get closer to the right answer.

The proliferation of data analyses used to support dubious claims, especially by politicians and industry, has led to a widespread skepticism of data analysis in general. In one of my other undergraduate courses, I devote an entire lecture to subtle and not-so-subtle distortions of the truth by politicians on both sides of the aisle. Nevertheless, while I admit that I'm sympathetic to the famous sentiment that Mark Twain attributed to Benjamin Disraeli—"There are three kinds of lies: lies, damned lies, and statistics"—I think the eminent statistician Fred Mosteller was correct when he replied, "It's easy to lie with statistics, but it's easier to lie without them." Explaining human behavior is an inherently difficult and controversial business, and the best hope we have of succeeding is to hold our claims up to the data and see how well they're supported.

Objective data analysis is especially important when we're trying to answer questions like whether war is in decline, because human intuition can lead us far astray. People are more likely to remember salient events like wars, so we're more likely to weight such events more heavily than we should when we're trying to assess historical trends—almost everyone would take the Iraq War into account when tallying conflicts, but few would remember to include the Chaco War or the Battles of Khalkhyn Gol. By the same token, we tend to ignore non-events, so we don't notice the absence of conflict nearly as much as we should. And history has a virtually limitless catalog of both kindnesses and atrocities to draw from, so reasoning by anecdote can't be trusted. For all of those reasons, we really need concrete data to help us understand whether war is in decline.

I think most people understand the idea of using data to understand human behavior fairly well. What many people don't understand, in my experience, is the fact that any science is an iterative process involving many people. We all have our biases and our blind spots and our shortcomings, and the only guarantee that we'll continue to progress toward the right answer comes not just from presenting our ideas to a community of skeptical scholars but in providing them with the means to evaluate those ideas independently. Being a scholar means not being committed to any one answer but, instead, being committed to the process by which questions are answered. It means exposing your reasoning and your data to public scrutiny and welcoming useful disagreement. It means welcoming a better answer, even (especially!) if it means that you yourself were wrong. The process is *only* credible if scholars both critique one another and care more about finding the right answer than about promoting their own.

Toward that end, I've done everything I can to enable other scholars to shoot holes in my arguments and build better ones. I've made the process by which I arrived at my conclusions as transparent as possible. That way, any of my colleagues who disagree can do things differently and see whether they get different results. For the same reason, all the data used in the analysis along with the command files used to execute it, all of which were written in the open-source R statistical language, are publicly available on my Dataverse, which can be found online at http://thedata.harvard.edu/dvn/dv/Braumoeller. If anyone succeeds in coming up with an answer that's better than the one I've laid out here, I'll happily write a blurb for his or her book.

Finally, because this is a question that has prompted widespread interest, I've gone to pains to explore it in a style that will be accessible to everyone. A large and growing number of data nerds—Nate Silver of fivethirtyeight.com, Professors Paul Krugman of the *New York Times* and Steven Levitt of *Freakonomics*, Christian Rudder of OKCupid, the colleagues of mine who contribute to The Monkey Cage column at the *Washington Post,* and others—have proven that it's possible both to understand data and statistics and to write in plain English. I hope to live up to their example.

Bear F. Braumoeller
Columbus, Ohio
April 2019

ONLY THE DEAD

PART I
Only the Dead

1

Introduction

The Spread of Peace and the Spread of War

World War II was the bloodiest war in human history. It killed somewhere on the order of 65 million people—a total that represents nearly 3% of the entire population of the globe. Overshadowed by World War II, the war in Vietnam killed an estimated 4.2 million people, a number roughly equal to the number of people killed during the entirety of the Napoleonic Wars. The Korean War killed roughly 3 million, but the starvation and repression that followed in its wake may have killed another 12 million.[1]

And that is far from all. Political scientist R. J. Rummel has compiled a list of what he calls "democides"—murders by government—that includes individuals killed by China (77 million, including the Great Leap Forward), the Soviet Union (61.9 million), Nazi Germany (20 million), and other brutal regimes in the twentieth century. His total for the century is a breathtaking 212 million deaths.[2] By any standard, that's a staggering toll. And it doesn't count homicides, injuries, attacks, and other forms of violence by humans against humans. Even today, as this book is being written, 68.5 million people—one person out of every 112 worldwide—has been displaced from his or her home as a result of conflict, persecution, or violence.[3]

In short, within the span of a single human lifetime the globe has witnessed wars, deaths, and massacres that most of us genuinely cannot imagine, and there's plenty of evidence that people still suffer the effects of organized violence on an enormous scale. Yet in recent years, some very smart and well-respected commentators[4] have argued that, *contra* popular perception, the human propensity to engage in violence in general, and in international armed conflict in particular, is on the decline and has been for decades or centuries. How can this possibly be the case?

The argument of this book is that it can't. As much as we'd love to believe that war has been on the decline in recent centuries, the evidence just doesn't support that claim. War has not been getting less deadly, whether you measure it in absolute terms or relative to the populations of the combatants.[5] The causes of war have not been getting systematically less potent. And although the worldwide rate of conflict initiation did decrease at the end of the Cold War, it had actually been steadily *increasing* prior to the 1990s.

That's not to say that there haven't been islands of peace here and there in the international system. There have, and they've been big enough and peaceful enough that they're worth explaining. In fact, these islands of peace have become more common and more persistent over the last two centuries.

How can that be the case? How can we see, simultaneously, the spread of peace and the spread of war? I argue that *international orders,* like the Concert of Europe and the postwar Western liberal order, tend to keep the peace among their members. International order is not by any means an unmitigated good when it comes to peace, however: International orders make military conflict with third parties more likely, both by creating or exacerbating conflicts of interest and by undermining the negotiations that could prevent them from producing war. On balance, the equation still seems to favor peace when only a single international order exists, but the coexistence of multiple orders is a recipe for conflict. The clearest example of such a dynamic is the plague of proxy wars between the West and the Communist bloc that spread around the globe during the Cold War.

This complex relationship between international order and international conflict is what drives the simultaneous spread of war and peace. It has both positive and negative implications for the short-term future of the international system. On the one hand, the demise (or, as seems more likely [Luce, 2017] retrenchment) of the Western

liberal order might make some kinds of conflict less likely. It would probably ease tensions with other major powers, like Russia and China, that find the power and ambition of the present-day Western order to be troubling, and it might well reduce the West's appetite for intervention. On the other hand, a weakened Western liberal order might allow frictions to emerge among its members and reduce their ability to prevent conflict elsewhere. The rise of alternative orders based on authoritarian principles, were it to occur, would likely produce the worst of all possible combinations from the perspective of peace: two large and powerful international orders, grounded in conflicting principles of legitimacy, with the ability to contest one another's influence anywhere on the planet.

As I argue in chapter 7, this tension between the two sides of international order is difficult if not impossible to escape. The same aspects of international orders that dampen conflict among their members create tensions between their members and the rest of the world, and particularly among members of rival international orders. What this means for humanity is that efforts to reduce conflict by creating international order may actually make matters worse when it comes to conflicts with third parties. The result can be an improvement over the status quo—even a very considerable one. But the inherent coupling of internal peace and external conflict make it very unlikely that the scourge of war can ever be fully eliminated. As the Spanish-American essayist and philosopher George Santayana put it in the wake of World War I, "Only the dead have seen the end of war."[6]

The Arguments

The three main arguments in the decline-of-war literature are worth comparing and contrasting, as each has different implications. Professor John Mueller argues that the institution of war, like the institutions of dueling and slavery before it, is simply going out of style. In his first book on the subject (Mueller, 1989), he argued that this change in attitudes was taking place among major, or developed, countries. For this reason, he argued that nuclear weapons were "essentially irrelevant"—the normative prohibition against war already guaranteed peace among nuclear states.[7] In his second (Mueller, 2007), he argued that the normative prohibition against war, which was largely the product of the horrors of World War I, has been spreading from developed countries to the rest of the world ever since.

Professor Joshua Goldstein argues that the decrease in warfare is quite recent, and he does not argue that it is irreversible. He explicitly does not argue that the evolution of civilization has produced increasingly strong norms of nonviolence, largely because if this were the case we would see a gradual decline in violence over time. But according to him, we see "a long series of ups and downs culminating in the horrific World Wars" (Goldstein, 2011, 42). Goldstein does argue that the end of the Cold War helps explain the reduction in violence after 1989, but for the most part he points to peacekeeping as the key causal variable: He argues that peacekeeping started having an impact on war in 1945 and accelerated after 1989, mostly due to the effect that peacekeeping has on the durability of settlements (pp. 43–44). The drop (p. 238) that he demonstrates in battle-related deaths per year is substantial, but it is also very recent, really only since the mid-1990s. Finally, he explicitly notes (p. 238) that "there is no guarantee that today's lull will last."

Professor Steven Pinker (2011*a*) argues that there has been a gradual decline in violence in general over the course of millennia, that the deadliness of international conflict has decreased since 1945, and that we now live in what is arguably the most peaceful period in history. He argues that three overlapping and somewhat irregular processes have driven this decline: the pacification process (humanity's transition from hunter/gatherer societies to agricultural civilizations), the civilizing process (the gradual strengthening of domestic authority and growth of commerce), and the humanitarian revolution (the expansion of empathy that resulted in the abolition of slavery, dueling, and cruel physical punishment and an increase in pacifism). In all, he (2011*a*, 192) argues that there are four components to the long-term trends in interstate warfare: "No cycles. A big dose of randomness. An escalation, recently reversed, in the destructiveness of war. Declines in every other dimension of war, and thus in interstate war as a whole." While he (2011*a*, 251–255, 361–362, 377) goes to some lengths to emphasize that he is not necessarily predicting that large wars will never happen again, a 2014 interview with NPR's Arun Rath made it clear that he thought a return to violence to be exceptionally unlikely. "How likely is it that we're going to start throwing virgins into volcanoes to get good weather or that you're going to have a return of slave markets to New Orleans?" he asked Rath. "I think pretty unlikely" (Rath, 2014).

While these authors differ in terms of the reasons that they posit for a decline in warfare, the exact aspects of warfare that they think have declined, and the duration of that decline, it is possible to discern three core arguments that make up the decline-of-war thesis, each of which is emphasized to different degrees by different authors:

1. Whether due to the spread of peaceful norms or the increased efficacy of peacekeeping, the causes of war are losing their potency. The same stimulus that would have produced war one hundred or two hundred years ago is considerably less likely to produce it today.
2. Because the causes of war are losing their potency, wars are initiated less often than they were decades or centuries ago.
3. When war does happen in the post–World War II era, its deadliness has decreased relative to previous periods due to the existence of widespread norms of nonviolence.

Evaluating the Arguments

Evaluating the decline-of-war thesis is the first goal of this book. It will not address the larger subject of violence in general, for two reasons. First, untangling the question of whether international conflict is on the decline is complicated enough to warrant a book in and of itself. Second, other scholars in other fields will be far better suited to address those questions than I am. As a student of international security, I feel reasonably confident in taking on the question of whether or not warfare is in decline. I feel far less confident in making such assessments about, say, trends in criminal behavior. That's not to say that I don't have opinions on the subject, but people who study it for a living will have far more informed perspectives than mine.

The question of whether or not warfare is in decline is an important one to answer, for a few reasons. Many of us want to know what sort of world we live in—and what sort of world we'll be leaving to our children. Can they expect to see tens or hundreds of millions of people die as the result of organized state violence in their lifetimes, as we have? Or are they really entering a golden era of peace and prosperity, conditional on the continued spread of Enlightenment ideas? What do the answers tell us about humanity more generally? Are we increasingly civilized, disinclined to use force to settle our arguments, or do we remain in a primitive state of nature—as Tennyson put it, "red in tooth and claw"?

A more immediate reason for writing this book is that the decline-of-war literature might actually have an impact on foreign policy. As I write this, the United States is debating how to deal with Russia's newly assertive foreign policy in central Europe, Americans are becoming ever more concerned about terrorist groups in the Middle East and nuclear weapons in North Korea, and scholars and policymakers who take the longer view are wondering just what American relations with China will look like a decade or two down the road. If warfare is indeed on the decline, we can afford to react minimally to each of these potential threats in the hopes of avoiding a lethal spiral of escalating conflict. If it is not on the decline, however, our passivity might cost us dearly. As my colleague Jennifer Mitzen (2013*a*, 525) writes in a review of Pinker's book titled "The Irony of Pinkerism,"

> [t]here is something deeply unsettling about the argument of this book. While I began reading without either smug comfort in my own circumstances or indifference to the violence that remains, by Pinker's final sentence on page 696 it was impossible to muster any other reaction. Indeed, I want to suggest that Pinker's book produces the type of reaction that conceivably could stop this important trend dead in its tracks. A world of elites and foreign policy decision makers well-schooled by Pinker in the causes of the decline in violence would be a world unmotivated to work to sustain it.

This concern is, I think, very well founded. Humanity seems most inclined to work for international peace when the danger of war is most apparent—typically, in the aftermath of huge wars like the Napoleonic Wars and the two World Wars. Fostering the belief that war poses no threat is a good way to convince people not to prepare for it. If warfare is not in decline (and perhaps even if it is), that sort of complacency will likely make things worse rather than better.

And I think there are some very good reasons to be skeptical of the claim that warfare is in decline. There are enough of them, in fact, that I broke them up across two chapters, which make up Part II of the book. In chapter 2, I review the data that have been used to support the decline-of-war thesis and conclude that they're just not very compelling. In chapter 3, I discuss some of the explanations offered for a decline in warfare and argue that those, too, fail to stand up to close scrutiny. This may seem like a substantial amount of space to devote to criticizing an argument, and it is. But this argument is pretty widely known, and a lot of people won't be inclined to change their minds about it. I've tried to provide an ample list of reasons to do so.

That said, the existence of substantial flaws in the evidence and arguments used to support the decline-of-war thesis doesn't necessarily mean that it's wrong. It just means that the question hasn't been answered. I do my best to answer it in Part III. In chapters 4–6, I assess trends over time in three key quantities: the rate of international conflict initiation, the deadliness of warfare, and the potency of the causes of war. I find that the nascent conventional wisdom on the subject—that warfare has been in decline for decades or centuries, despite appearances—is just wrong. We do see a decrease in the rate of conflict initiation at the end of the Cold War, but that's about the only good news. Other than that, for the last two hundred years at least, I can find no general downward trend in the incidence or deadliness of warfare. If anything, the opposite is true. I show that the same conclusion holds pretty much regardless of how we measure international conflict and the deadliness of warfare. As to the question of whether the same circumstances produce less warlike responses today than they did fifty or a hundred years ago, I find that for the most part they do not. There is significant variation in the potency of the causes of conflict both within and across the studies that I examine. Some are in fact becoming less potent. But about an equal number are becoming more potent.

War, Chance, and Numbers

One of the biggest contributions of this book is that it brings statistical inference to bear on the question of the decline of war. I make every effort to do so in plain English, conveying enough information to interpret the figures and leaving the rest to the Appendix. Even so, some readers might wonder why the tests are necessary. Can't we simply look at trend lines and see what's going on?

Unfortunately, we can't. Following every uptick or downturn in a trend line can be deceiving because, given how we believe international conflict works, it's reasonable to expect the timing of conflict onset and the deadliness of war to be at least somewhat random. A lot of people find it difficult to believe that there might be some random component to a human behavior that's as deliberate as the initiation of international conflict: War is generally the result of a very serious decision, and we really don't like to believe that such decisions contain an element of randomness.

There's good reason to believe that they do, however. While we've made good progress in past decades in understanding the circumstances that make conflict more likely, we can probably never know exactly when a conflict will break out, a leader will be assassinated, or a country will collapse into violence. It is virtually impossible to know when an unforeseen event will render an underlying conflict salient enough to focus the attention of the parties involved on resolving it, and even if we knew that there is no way to know whether negotiations will result in a settlement or an armed conflict. In a handful of cases, one country recognizes that negotiations would be either fruitless or counterproductive and forgoes them in favor of simply attacking without warning. We certainly can't know when that will happen: The whole point of a surprise attack is that it's a surprise. And once a war begins, it's very hard to know how many people will have died before the shooting stops.

From this point of view, international conflicts occur primarily because underlying causes of tension between two countries increase the probability that a conflict will be initiated. Once that tension exists, the countries involved essentially "roll the dice" every month or every year. Depending on how serious the tensions between them are, they might fight only if they roll snake-eyes, or they might fight if the total of the two dice is four or below, or... well, you get the idea. The underlying probability of conflict may not change much from year to year, but the outcome—whether or not we actually see a conflict—does.

The combination of long-term issues or circumstances that predictably make war more likely and short-term circumstances that make its timing very difficult to predict means that students of international conflict are a lot like seismologists. We can map out fault lines and demonstrate that some parts of the world are more dangerous than others, but we can't tell you that your part of the world will be really, really dangerous next Tuesday.

That fact makes testing arguments about wars and conflicts challenging. Because we cannot predict the timing of a single conflict initiation, we cannot predict exactly how many will occur over a given period of time. All we can do, really, is describe the probability distribution that is generated by the underlying causes of tension. Lewis Fry Richardson, for example, demonstrated in his thoughtful book *The Statistics of Deadly Quarrels* that conflict onset follows a Poisson distribution, which describes the number of events that we should expect to see across many periods of the same length (days, for example) if those events occur at random but with a constant underlying rate.[8] The fact that conflict onset follows a statistical distribution does not mean that wars happen entirely at random or that no deeper cause exists. What it means is that the causes of war do not produce war with certainty: they increase the probability of war, and circumstance does the rest.

The years leading up to World War I in Europe provide a good example of this process.[9] Following the resignation of German Chancellor Otto von Bismarck and especially the formation of the Franco-Russian Alliance in 1894, and in the context of the clash of imperial ambitions abroad and nationalist and imperial ambitions in the Balkans, European politics became increasingly polarized and militarized—a "powder keg" waiting to explode. There were at least four crises in the years leading up to World War I that could have produced the spark needed to ignite this volatile situation: the First Moroccan Crisis (1905–06), in which France and Germany clashed over Morocco; the Bosnian Crisis (1908–09), in which Russia and Serbia were prepared to go to war over the Austro-Hungarian annexation of Bosnia and Herzegovina; the Agadir (or "Second Moroccan") Crisis of 1911, in which Britain threatened war to forestall German aggression against France in Morocco; and the Balkan Wars of 1912–13, in which the Balkan clients of the major powers

FIGURE I.1 A 1906 political cartoon by John Scott Clubb depicting the state of constant tension among the major powers in Europe.
Source: Library of Congress.

fought the Ottoman Empire and one another. Given the number of countries involved, the intensity of their interests, the mistrust inspired by the Anglo-German arms race, and the ubiquity of misperception, it is not difficult to imagine any one of these crises evolving into a war that would engulf the continent.

Moreover, the event that actually did start the war very nearly didn't happen. No fewer than six assassins lay in wait with guns and bombs along Archduke Franz Ferdinand's motorcade route in Sarajevo on June 28, 1914. When the time came, the first two failed to act. The third, Nedeljko Čabrinović, threw a bomb that bounced off of the Archduke's car and detonated underneath the car behind it. The Archduke's driver sped past the remaining assassins, depriving them of their opportunity. One of them, Gavrilo Princip, took up a position along the motorcade's next planned route, but the Archduke and the Governor of Bosnia, Oskar Potiorek, wisely opted to change plans precisely to avoid the possibility of encountering more assassins. Unfortunately, the route that they chose ran along the Appel Quay, as had the original route. Even more unfortunately, the drivers never got the message about the change in plans. When they began to turn off of Appel Quay up Franz Joseph Street along the original route, they were stopped by Potiorek and told to turn back. As the Archduke's car sat motionless, Princip—who, as fate would have it, had chosen to wait for them at that exact intersection—stepped up to the car and assassinated the Archduke and his wife.[10]

In short, it was not difficult to know that Europe in the early 1900s was a pretty dangerous place and that war among the major powers was a real possibility. It was effectively impossible, however, to predict that the Bosnian Crisis would not start such a war but that the assassination of the heir to the Austro-Hungarian throne would—especially when that assassination depended on the Archduke's driver making a wrong turn at the corner where a young Serb assassin was waiting.

World War I is far from unique in having been an issue-packed powder keg awaiting some essentially random spark. In fact, a fair number of wars and crises have been named after the idiosyncratic events that set them off. A brief sample:

- The underlying causes of **the Soccer War** between Honduras and El Salvador were tensions having to do with emigration and land scarcity, but the war itself was sparked by escalating riots surrounding the 1970 FIFA World Cup qualifier.
- **The Pastry War** of 1838–39 had its roots in economic disputes between Mexico and France but was set off by the looting of a French pastry chef's shop in Mexico City.
- **The War of Jenkins' Ear** was also fought over economic issues, but the proximate cause was an incident in which a British captain, Robert Jenkins, was accused of smuggling by the captain of a Spanish coast guard boat and had his left ear cut off as a warning to the king.

- **The Postage Stamp Crisis** arose from an underlying territorial dispute between Nicaragua and Honduras. Troops were mobilized after Nicaragua printed a stamp depicting the disputed territory as being part of Nicaragua.[11]

I don't mean to imply that, if these sparks had not occurred, the wars and crises that ensued would not have happened. Far from it, in fact. The assassination of the Archduke was an isolated and tragic incident until Austria-Hungary used it to create an excuse for war with Serbia. Captain Jenkins's ear was separated from the rest of him in 1731, but the war that bears its name did not begin until 1739, when British trade disputes with Spain moved Parliament to sudden if belated outrage over the captain's maiming. It may be, as McGeorge Bundy said when he was asked why the 1965 Viet Cong attack on Camp Holloway, near the city of Pleiku in South Vietnam, had provoked the United States to escalate by attacking North Vietnam, that "Pleikus are like streetcars"—that is, that small provocations occur fairly often, and any of them could serve as a rationale for the use of force.[12] The larger point is simply this: Even if Pleikus really are like streetcars, one can never know exactly when they'll come along.

Precisely because we don't know when exactly they'll come along, we can't draw any meaningful conclusions based on year-to-year changes in the rate of conflict onset or the number of battle deaths.[13] Those numbers are going to fluctuate to some degree just by chance. And since there's a random component to how big wars get—a very big random component, it turns out, as I'll discuss in chapter 3—we can't draw any meaningful conclusions just by eyeballing trends in battle deaths either.

That's why we need formal tests that are designed to separate the underlying rate of conflict initiation and the deadliness of wars from the "noise" of chance variation. Without such tests, it's very difficult to say anything meaningful about whether the world is getting more or less peaceful.

Explaining Systemic Trends in Warfare

Although these tests indicate no general decline in the rate of conflict initiation, chapter 4 does conclude that there is some systematic variation over time. That fact raises an interesting question: if we're not seeing a general trend toward a less bellicose world, what *is* going on? Why is international conflict more common in some periods than it is in others?

Answering that question is the second goal of this book, which I take on in Part IV (chapters 7–8). I find the answer in patterns of *international order*, which produce "islands of peace" throughout the history of the last two hundred years. Unfortunately, while those islands tend to foster peace among their inhabitants, they are prone to go to war with one another—as the Western allies and the Soviet Union did throughout the Cold War, and as the Soviets, Nazis, Japanese, and Western powers did in the interwar period. The double-edged nature of international order greatly complicates the relationship between international order and international conflict and undermines the simple idea that more (or better) order automatically leads to more peace.

To understand what I mean by international order, first consider a somewhat stylized account of the origins of a more familiar phenomenon: domestic political order, or governance. Probably the best-known account of domestic order argues that it happens when people get tired of killing each other and taking each other's stuff and decide that life would be better if they could all agree to appoint someone to protect their lives and property.[14] They strike what amounts to a bargain with each other, in which they all grant governing authority to a sovereign in exchange for protection. The minimal justification for this arrangement is peace for the governed.

In subordinating themselves to a larger political order, however, the governed soon discover that they have enabled warfare on a scale that they had not previously imagined. Whereas previously individuals and small gangs had fought one another, now the protection of the group requires mobilization of a significant fraction of the population for a common goal—usually, killing a significant fraction of some *other* group. Rather than disappearing, violence becomes organized violence, and as this feudal arrangement evolves into a modern state organized violence becomes war.

The essence of this hierarchical two-step was captured succinctly and brilliantly by the sociologist Charles Tilly (1975, 42), who wrote that "[w]ar made the state and the state made war." In other words, warfare among individuals prompted those individuals to form the state, and once the state was formed it gained the ability to wage war on a larger scale.

International order, fundamentally, is much the same thing as domestic political order—a set of expectations about behavior and outcomes that allow a society to function as a society. It is not *government* in the sense that we think of domestic government, with citizenship and flags and perhaps elections, but it is *governance* in the sense that states willingly submit to some form of broader international authority. Examples, from the post-Napoleonic Concert of Europe to the postwar liberal international order, are not hard to find in modern history, though they are relatively few in number.

My answer to the question of what drives systemic changes in the rate of international conflict is simply this: Professor Tilly's argument about domestic political orders applies to international orders as well. War makes international order, as Professor G. John Ikenberry (2001) pointed out: The aftermaths of the great wars of the nineteenth and twentieth centuries have witnessed concerted efforts to create and implement different forms of international order, from the Concert of Europe to the postwar Western liberal international order. And while isolated international orders on balance tend to promote peace, multiple orders make war—especially when they are founded on incompatible principles of legitimacy. Accordingly, what we see in the historical record and the data, both of which I examine in chapter 8, isn't so much a constant movement toward a less conflictual world but rather an irregular series of shifts from more conflictual periods to less conflictual ones and vice versa, based largely on which international orders existed at the time. Chapter 9 summarizes the results and offers some concluding thoughts on the nature of war.

The implications of the book for our understanding of peace and war are tragic. We do seem to have invented a way to create islands of peace in the international system, and that invention is an innovation of monumental importance. The absence of conflict within those islands is very real. But we are not often inclined to do so unless history has just given us a compelling reason. Too often, that reason is an enormous war. In the absence of such a war, we seem disinclined to make any sustained effort, sacrifice our autonomy, or compromise our national values to come together and create a new, lasting security architecture that would further enhance the prospects for peace. The tragedy is that, while we've learned a lot about how to make peace, we're not especially inclined to do it at present. It is as though humanity, having been given the gift of fire by Prometheus, used it to make a lovely bonfire, watched the fire burn down to embers, shrugged, and went to bed.

PART II
Reasons for Skepticism

2

Reasons for Skepticism, Part I: Data

AS I WILL SHOW later in the book, the arguments that make up the decline-of-war thesis are readily amenable to data analysis. Both Goldstein's and Mueller's claims suggest a decline in the rate of conflict initiation, though the time frames differ a bit. Pinker's argument suggests a much longer-term decline in every dimension of warfare—judging by some of his data analyses, a decline dating back to the fifteenth or sixteenth century. It's possible that a "big dose of randomness" might obscure these trends, but the tools of statistical inference are designed precisely to give us useful inferences in the face of lots of randomness.

Before I analyze the data, though, I want to take some time to explore the existing findings that support of the decline-of-war thesis and explain why I don't find them compelling. A lot of people have been convinced by the decline-of-war thesis, and I owe it to those people to explain why I'm not. Most of my reasons for doubting the decline-of-war thesis have to do with the evidence offered in support of specific claims. As I mentioned earlier in the book, the more I read about the decline-of-war thesis, the more I found myself filling margins with questions and objections. Exploring all the questionable findings or claims that inspired those annotations would make readers' eyes glaze over, especially given that some are abstract and difficult to resolve one way or the other. But I've included objections to five of the most important data-based claims that support the decline-of-war thesis in order to illustrate my reasons for believing that the thesis deserves further exploration.

Trends in Conflict Initiation

One of the most compelling graphs in Professor Pinker's book about the decline of war is a graph in chapter 5, Figure 5-17, which shows a general downward trend in the number of conflicts per year that were initiated in Europe since the 1400s. I have reproduced it here as Figure 2.1. Out of curiosity, I dug out the conference paper that Pinker cites as the source of the data (Brecke, 1999), and I found a lot of detail about how the data—Professor Peter Brecke's "conflict catalog"—were compiled. I also found two figures—Figures 4 and 5, the number of conflicts per decade in Western and Eastern Europe, respectively—that map well onto Pinker's Figure 5-17, though Brecke aggregated the data into decades while Pinker chose twenty-five-year periods. The trend looks very impressive either way—it appears as though Europe, at least, has become a much more peaceful place than it once was.

The main problem with this graph, of course, is the problem that my students pointed out: The number of political entities is not constant over time. Europe in 1400 was a dazzling patchwork of independent and quasi-independent polities. As their number decreased over time, the number of opportunities to fight decreased as well. Simply by looking at this graph, we can't really tell whether the probability that one country will attack another has increased, decreased, or stayed the same.

That point underscores a key distinction that will be repeated throughout this book: the distinction between the *frequency* of conflict and the *rate* of conflict. The *frequency* of conflict initiation is the number of occurrences of a specific conflict event in a given time period. Here, the measure is the number of conflicts (armed clashes resulting in at least thirty-two deaths) per year in Europe, averaged over twenty-five-year periods. The *rate* of conflict initiation, on the other hand, is the frequency of conflict initiation divided by the number of opportunities to initiate conflict. This number answers the most basic question we have about states: given the opportunity to fight, how often do they do so? The number of conflict opportunities is a bit hard to gauge in the context of these data, because interstate wars are mixed in with civil wars, genocides, riots, and insurrections—another source of confusion!—but it's safe to say that, as the number of states decreases, the number of opportunities to engage in interstate and civil war should decrease as well.

Unfortunately, the measure in the graphs—conflict initiations per decade or twenty-five-year period—doesn't really include any information about the number of conflict opportunities per year, so it's impossible to determine just from the information presented whether the rate of conflict is increasing or decreasing. If the number of conflict opportunities hasn't changed very much, it might be decreasing; if the number of conflict opportunities has decreased dramatically, the rate of conflict might even be increasing. We really don't know.[1]

Another issue has to do with the geographic scope covered by the data in the graph. Pinker argues that conflict is in decline worldwide, but the data in the figure only cover Europe—a fact that he justifies by arguing that Europe is both data-rich and consequential. That seems fair, especially if data from elsewhere are actually unavailable. But they're not. They're contained in the same dataset. As I mentioned above, Pinker's Figure 5-17 is derived from the data in Figures 4 and 5 of a conference paper. Figure 1 of that same paper, reproduced here as Figure 2.2, captures the frequency of conflict initiation worldwide.

This figure tells a dramatically different story than the first one does. Far from being the most peaceful time in our history, the twentieth century is surpassed only by the nineteenth in terms of the number of conflicts started. If there is a downward trend in the twentieth century—and that's a big "if"—it's not the culmination of a long

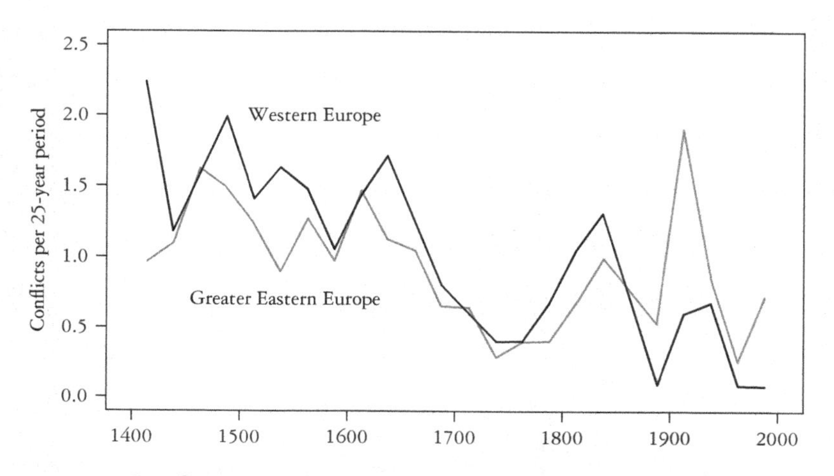

FIGURE 2.1 Conflicts per 25-year period in Europe.

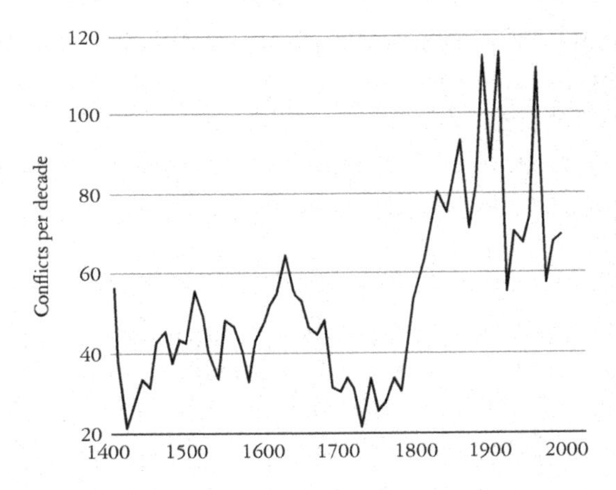

FIGURE 2.2 Conflicts per decade worldwide.

historical process of pacification but rather a modest decline from the most bellicose period in recorded history.

It's possible to make too much of these trends, of course. In fact, my own take is that neither of these graphs really tells us much of anything. There's too much going on—changes in population size, erratic and unpredictable changes in the odds that a violent incident will be recorded for posterity, and a transition from a feudal order prior to the Thirty Years' War to a system of sovereign states afterward, just to name a few[2]—for us to be able to glean much of anything about trends in conflict from either of these graphs. But if Pinker is right and Figure 2.1 really does contain useful information, Figure 2.2 should have been entered into evidence as well.

The "Long Peace"

The decline-of-war thesis seems implausible on the surface, if for no other reason that the twentieth century was shockingly bloody by any measure. The response of its proponents, essentially, is that one or both of the two World Wars in that century was a fluke—a statistical anomaly. Mueller (2004, 4), for example, calls World War II "a spectacular anachronism, fabricated almost single-handedly by history's supreme atavism, Adolf Hitler." Pinker (2011*a*, 208–9) also points to

Hitler, as well as to Gavrilo Princip, and argues that each was a necessary condition for his respective war.

The problem with this claim is that it's often possible to point to one person who, more than others, lit the fuse that led to war. The Napoleonic Wars are called that for a reason. That's not the same thing as arguing that individual was both necessary and sufficient for the onset of war—an extremely ambitious claim that is unlikely to withstand scrutiny.[3] World War II may have been begun by Hitler, but the ground was made fertile for him by widespread bitterness about German reparations. The Allies took these steps knowing full well that there was a risk of backlash: Although no one could have foreseen Hitler, *some* hypernationalist response leading to a Great Power war was hardly unlikely.

The decline-of-war theorists prefer to focus, not on the "anomalous" World Wars, but on the seventy-year period that followed them, during which a similar general war among Great Powers failed to materialize.[4] As historian John Lewis Gaddis (1986, 99–100), who coined the term "the Long Peace" to describe this period, points out,

> the post–World War II system of international relations, which nobody designed or even thought could last for very long, which was based not upon the dictates of morality and justice but rather upon an arbitrary and strikingly artificial division of the world into spheres of influence, and which incorporated within it some of the most bitter and persistent antagonisms short of war in modern history, has now survived twice as long as the far more carefully designed World War I settlement, has approximately equaled in longevity the great nineteenth century international systems of Metternich and Bismarck, and unlike those earlier systems after four decades of existence shows no perceptible signs of disintegration. It is, or ought to be, enough to make one think.

The problem with emphasizing the importance of the Long Peace while writing World War II off to chance is that doing so attributes the occurrence of war to chance without exploring the role of chance in the occurrence of peace. In fact, most scholars see the process that produces both outcomes as being irreducibly probabilistic. One can cite any number of peaceful periods that could just as easily have exploded into war. During the Cuban Missile Crisis, for example, the United States conveyed its intention to use practice depth charges to force Soviet submarines near the quarantine line to surface, but the Soviet Union failed to inform the four Foxtrot-class submarines in the area of that fact. The Soviets also failed to communicate to the United States the fact that their submarines were armed with nuclear-tipped torpedoes. On October 27, a trapped Foxtrot submarine very nearly used its nuclear weapon against the American Navy: doing so required consensus among the captain, the second in command, and the political officer, and only one of the three—Vasili Arkhipov—refused to endorse a nuclear response.[5]

If we give the role that chance plays the same credit in peace that we give it in war, we can take a rough stab at assessing which of these periods—the two World Wars, or the seventy years that followed—was more anomalous. If we look at the record of Great Power wars over the past five centuries (Levy, 1983; Goldstein, 1988, 146), we find an average of about two per century prior to the twentieth century. Having two such wars in the twentieth century, therefore, is far from anomalous; it's more or less what we'd expect given what we've seen in the past. More surprising, perhaps, is the fact that the seventy years of peace that have passed since the end of World War II are hardly anomalous either.

To see this point, let's take the probability of a systemic war breaking out in any given year, based on historical precedent, to be 0.02.[6] We can calculate the probability of seeing a given number of wars over a given period of time by using the binomial distribution—the same distribution used by introductory statistics students to calculate the probability of observing "heads" in five successive flips of a coin. Here, we can use it to get a sense of just how unlikely a Long Peace of seventy years' duration really is.[7]

Figure 2.3 uses the binomial distribution to compare the probability of observing no wars in the seventy years since World War II with the probabilities of observing one, two, three, or more wars in the same time

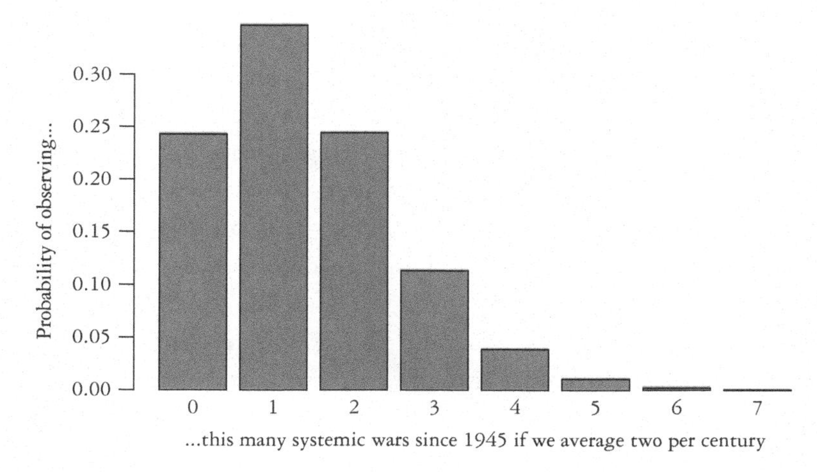

FIGURE 2.3 The entirely probable Long Peace.

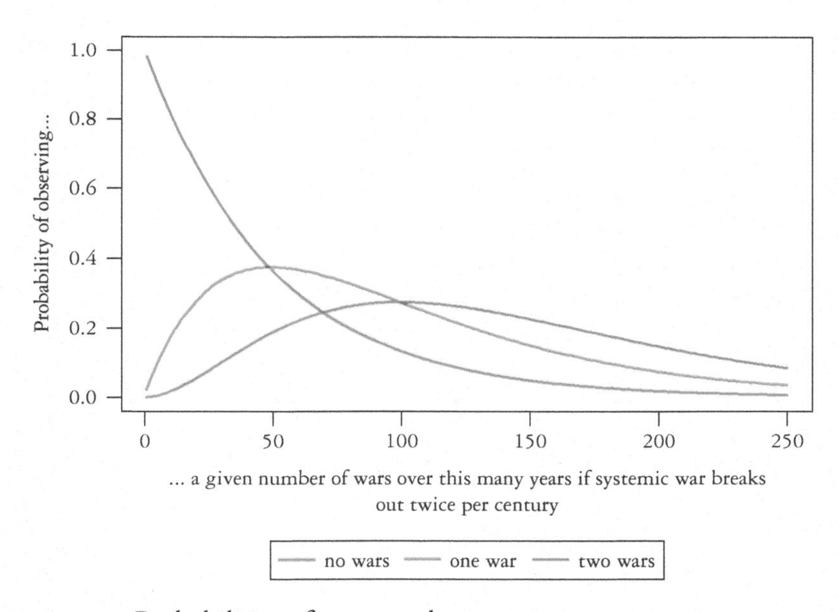

FIGURE 2.4 Probabilities of peace and war over time.

period, assuming that the probability of systemic war has not deviated from its historical average of two per century. As we can see, although the most likely outcome is a single war during that time period, an absence of wars is far from improbable.

Figure 2.4 takes a look at how these same probabilities change over the course of 250 years. As the figure demonstrates, for nearly fifty years, the absence of war is the most likely outcome. A single war then becomes more likely but not much more; and at no point are one or two wars much more likely than none at all.

These figures answer our question about the improbability of World Wars I and II, relative to the Long Peace. If there has been no change in the underlying propensity for systemic war over time, the probability of seeing two wars in the twentieth century—that is, the probability that two of the hundred one-year periods contained in the century will see the onset of general war—is 27.3%. By contrast, the probability of observing seven continuous decades of peace—zero wars in fourteen five-year trials, with a probability of "success" of 0.02—is 24.3%. In other words, the Long Peace is only slightly less likely to have happened by chance than the two World Wars.

How many years of peace would we have to see before we *could* conclude that the world has changed and that the long-term average of two general wars per century no longer applies? We can examine the credible range of our estimate of the probability of war over time to answer this question. These intervals delimit the plausible values of the probability of a Great Power war after the passage of a given number of years of peace. As Figure 2.5 shows, although the range of plausible values narrows quickly for the first half-century or so, it would still take *about 150 years* of uninterrupted peace for us to reject conclusively the claim that the underlying probability of systemic war remains unchanged.[8]

These are, admittedly, shirtsleeve calculations: they apply very simple math to get an approximate answer to a difficult problem.[9] In particular, they don't take into account the complexity of the process by which small wars escalate into large ones or the implications of changes in that process. That's not my goal at the moment (though it will be later on). My point is just that even these simple calculations should give pause to anyone who wants to claim that the seventy-year Long Peace, *in and of itself*, represents compelling evidence that the world has become a more peaceful place.

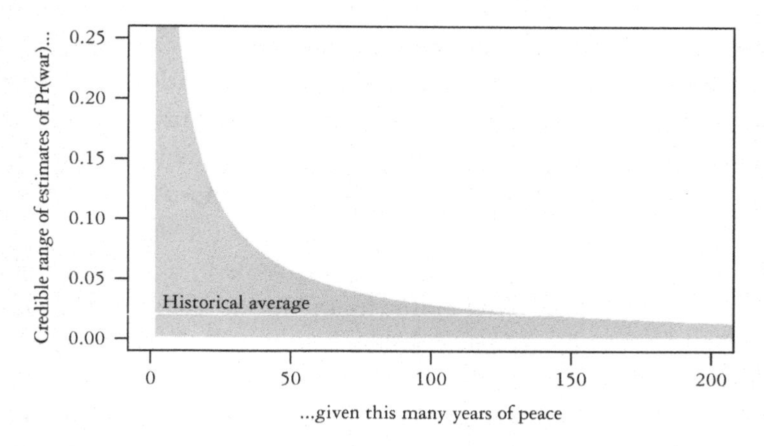

FIGURE 2.5 Credible estimates of the probability of war over time.

Attitudes About War

While these critiques call into question the larger claim that peace is breaking out worldwide, they don't really get at the narrower argument, made by professors Pinker and Mueller, that norms of peaceful behavior took root first in the most developed countries and spread outward from there. If that were the case, it's possible that an increase in violence among less developed countries might swamp a decline among developed ones, leaving us to conclude that no change has occurred.

The ideal way to assess this argument would be to look for norms of nonviolence among the populations of developed states. If Mueller and Pinker are correct, the populations of developed countries should be significantly less prone to advocate violence than the populations of undeveloped states. Despite the wealth of evidence presented to support the decline-of-war hypothesis and the prominence of the argument that nonviolent norms have spread, however, none of the authors spends much time on actual survey data.

That could be because the American public, at least, doesn't come across as especially pacifistic. A recent study of gun violence in movies (Bushman et al., 2013, 1014), for example, concluded that "violence in films has more than doubled since 1950, and gun violence in PG-13–rated films has more than tripled since 1985." Of more direct relevance, a recent survey by scholars from the College of William and Mary and the

University of Wisconsin gauged the American public's support for the use of force across eight different foreign policy scenarios. The scenarios that got the least enthusiastic responses, a Chinese invasion of Taiwan and a Russian invasion of either Estonia or Ukraine, still garnered 43% approval. The two scenarios that got the most favorable overall responses were combating Islamic State fighters (62%) and preventing Iran from developing a nuclear weapon (63%). In a summary of the findings, Professors Michael Horowitz and Idean Salehyan (2015) noted that "about 15% of the general public supported US military force no matter what the question."

That doesn't really answer the question of whether citizens of developed countries are in general less prone to advocate violent policies than citizens of less developed countries, of course. To do so, we would need to survey citizens in a broad set of both developed and less developed countries. Unfortunately, the overwhelming majority of surveys are national in scope. Very few are carried out in enough countries to make cross-national comparisons useful, and of those, hardly any contain survey questions that would help us to measure the spread of nonviolent norms.

I did, however, manage to find one question that both reflects norms of nonviolence and is asked recently of the citizens of a fair number of countries. Both the World Values Survey[10] and the German Marshall Fund of the United States' Transatlantic Trends survey[11] ask respondents to agree or disagree with the statement "Under some conditions, war is necessary to obtain justice." It is difficult, of course, to gauge attitudes about war without offering some proposed justification for it, just as it would be difficult to ask consumers whether they like spending money without telling them what they're spending it on. A question about going to war to defend the nation, for example, would probably get a far higher percentage of positive responses than a question about going to war for the sake of conquest would. I'm speculating a bit, of course, but having put in some time early in my career on survey research and question design I suspect that justice is probably as good a justification as any and better than most in a question like this. It's a plausible reason to fight but not necessarily a compelling one, and it's vague enough that respondents' sentiments about war, which are likely to be considerably more concrete than the idea of justice that the question brings to mind, probably drive a lot of the response.

Figure 2.6 captures the responses of representative samples of adults in fifty-seven countries to the question of whether war is sometimes necessary to obtain justice. The responses on the dotchart on the left are a snapshot in time: they took place during the World Values Survey's sixth wave, between 2010 and 2014.[12] The trend lines on the right are drawn from the Transatlantic Trends data and capture changes in aggregate responses to this question over time in thirteen of those fifty-seven countries; the gray zones in the background cover the range from 33% to 67%.

The first thing to note is that the trend lines do seem to trend downward a bit at first, though most that do so also trend upward subsequently. These trends are relatively minor in magnitude and plausibly reflect reactions to contemporary world events: 2004, at the beginning of the period, was a peak year for coalition casualties in Iraq, while 2009, the inflection point in many trend lines, was a peak year for coalition casualties in Afghanistan. These and other world events, along with the relatively short time span, make it difficult to tease out any meaningful long-term trends.

What is much clearer, however, is the comparison between developed (OECD) countries and less developed countries. The dotchart suggests that there's very little difference between the two in their citizens' responses to the question, and a little math confirms this impression: On average, 31.9% of respondents in OECD countries respond in the affirmative, while 31.6% of respondents in non-OECD countries do so. As this modest difference suggests, the responses of the two groups are statistically indistinguishable from one another.[13] The range of responses within each group is vastly greater than the average difference between them.

This is a fairly difficult set of findings to square with the claim that norms of nonviolence are spreading from the developed world to the less developed world. The data reveal neither a consistent downward trend in support for war among developed countries nor a meaningful difference in support for war between developed and less developed countries.

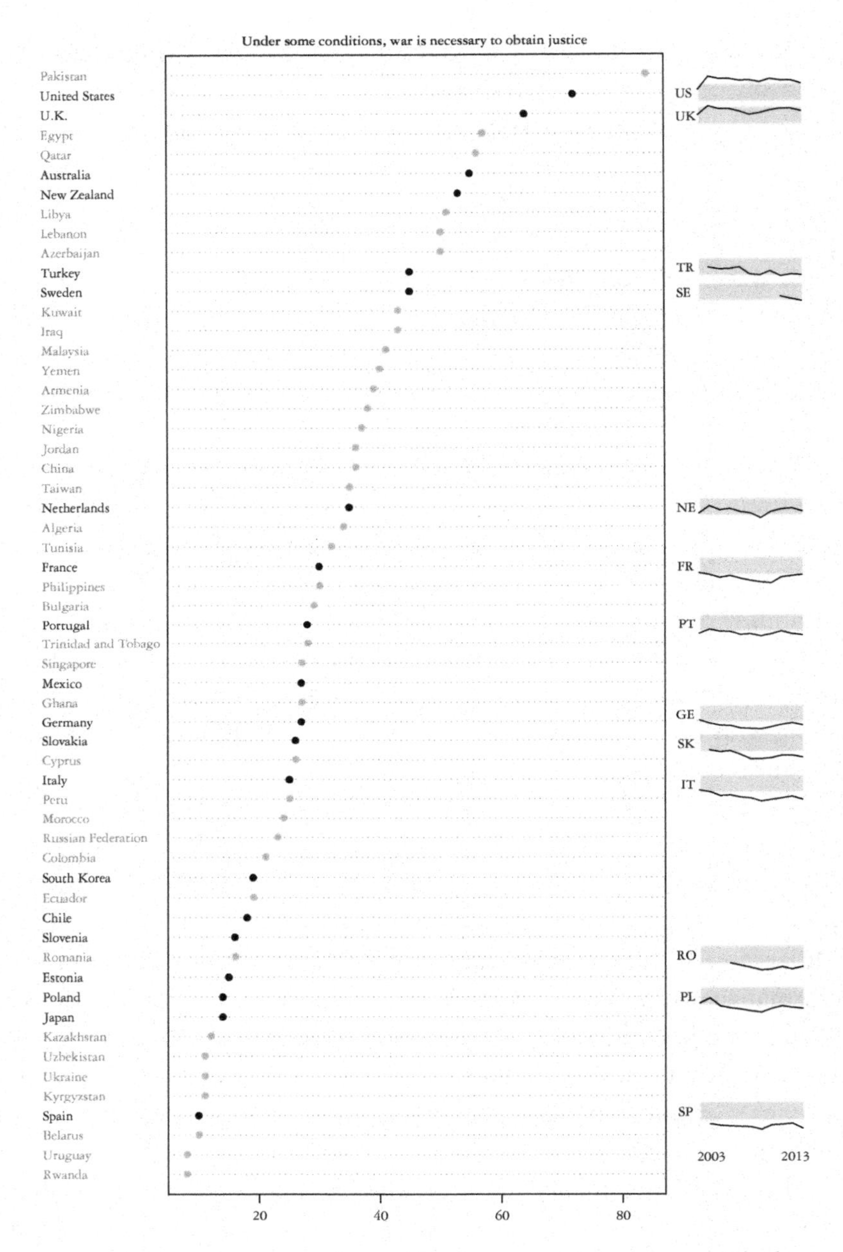

FIGURE 2.6 Percentage agreeing with statement, by country. Black text and dots indicate OECD countries. Sparklines on the right reflect 11-year trends (2003–2013) for a subset of states from the Transatlantic Trends data; the gray zone indicates the middle third of the range (33%–67%).

The Pacifying Effects of Civilization

Professors Pinker and Mueller both argue that civilization, in the form of the modern state, has a pacifying effect: Hobbes's "Leviathan" protects each of us from all of us and, in so doing, decreases the amount of violence in society. In Mueller's (2004, 176–177) case, this point takes the form of an argument that, in the present day, poor or weak governance significantly increases the probability of war. Pinker widens the scope of his inquiry considerably, to include both modern nonstate societies and prehistoric ones (Figure 2.7).

I think the argument that the modern state makes life peaceful for its inhabitants is probably right. A big part of the reason for having government in the first place is that it enforces your right not to be beaten, stabbed, or killed. But that generalization applies to *all* forms of political order, not just states—and it's not immediately obvious that states are best at ensuring the safety of the people who comprise them.

In fact, some states kill *a lot* of their own citizens. Mao's China and Stalin's Russia, both modern states, murdered their citizens in such staggering numbers that the state of nature would have been a most welcome alternative. At the same time, some nonstate societies turn out to be quite peaceful. So despite the fact that, at first blush, this is a reasonable claim, it's still worthwhile to examine the evidence. And that's where things get tricky.

First of all, the forensic anthropology that's required to assess cause of death based on skeletal remains is really hard to do and often frustratingly inexact. As a result, it's just not easy to know how representative or accurate forensic anthropologists' assessments of causes of death really are, especially when they lack access to the physical context in which death occurred and the soft tissues of the deceased. A recent essay by forensic scientists João Pinheiro, Eugénia Cunha, and Steven Symes (2015) in a book on skeletal trauma analysis presents two modern case studies in which conclusions regarding cause of death would have been dramatically incorrect had they been based solely on skeletal remains. Nor are these isolated cases: as the authors conclude, "the mistake of inferring too much from too little is one of the greatest problems in an anthropologist's routine analysis of bone injuries" (Pinheiro, Cunha, and Symes, 2015, 40). Add to that the

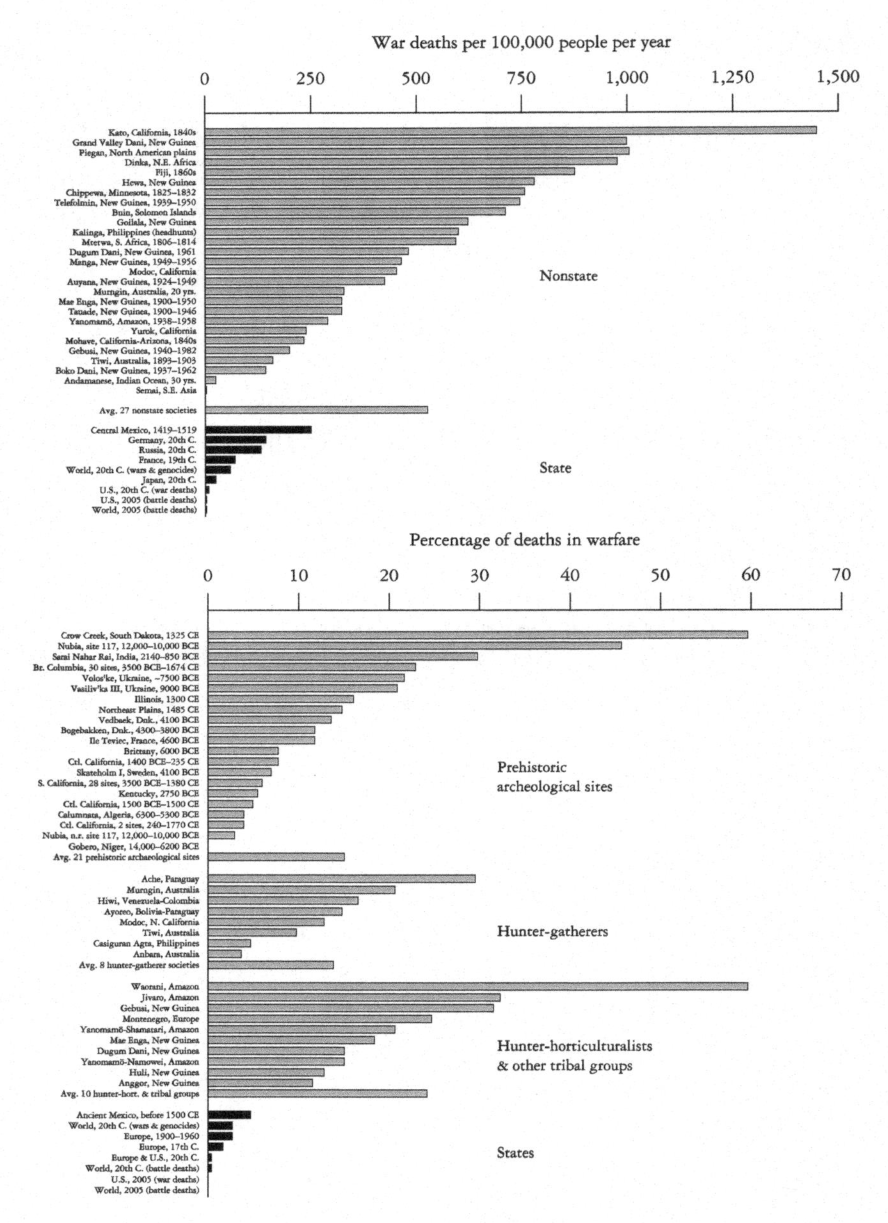

FIGURE 2.7 Mortality rates from warfare in prehistoric (top) and modern (bottom) nonstate societies, compared to mortality rates from warfare in state societies. Source: Pinker (2011*a*, 49, 53).

fact that skeletons are most likely not discovered at random (perhaps those who died in battle died nearer to one another than those who didn't or were buried in more prominent locations) and the physical evidence can be difficult to interpret (some societies bury people with weapons next to them, for example), and you begin to appreciate how the uncertainties in these estimates can begin to compound.

Largely due to the ambiguity of the available evidence, anthropologists themselves are divided on the issue of whether fighting in nonstate societies was rare or endemic—a fact that Pinker (2011*a*, 36) himself notes. The debate is far from settled. For example, in a recent volume devoted to this issue anthropologist R. Brian Ferguson (2013) dissects the available evidence in each of the prehistoric nonstate societies mentioned by Pinker. His chapter, which is dense with arguments about the implications of the locations of weapon fragments, the specifics of carbon dating, and descriptions of the relevant features of burial sites, makes for compelling reading. In the end, he (2013, 116, 126) concludes that the ubiquity-of-warfare argument "is utterly without empirical foundation" and that the cases are "cherry-picked"—a "selective compilation of highly unusual cases, grossly distorting war's antiquity and lethality." No mention is made, for example, of the relatively peaceful Hopi, Zuni, or Havasupai. Again, my intention is not to choose sides—both authors know quite a bit more about anthropology than I do!—but rather to point out that, based on the more systematic findings of the relevant academic community, there are solid reasons to be skeptical of the thesis that nonstate societies are systematically more warlike than state societies.

Moreover, even if we can trust the evidence, it might not signify what we think it does. Mueller's argument that weak states are more warlike is well supported by the evidence, but there is also strong evidence to suggest that the causal arrow runs in the opposite direction. Weak states rarely become weak out of the blue; more often, their weakness is precipitated by internal conflict or external intervention (Rotberg, 2004). And a high rate of death among modern tribes may simply reflect what we already know about many of them: that they were killed off in substantial numbers by supposedly more peaceful state societies. For example, perhaps a quarter of the Dinka—one of the tribes in Pinker's book whose members were most prone to violent death—were wiped out, not in conflicts with other tribes, but rather at the hands of Sudanese government forces during the Second Sudanese Civil War (1983–2005). Since we know that very few nonstate societies have prevailed or even held their own against modern states, it hardly seems surprising that their populations would have perished in greater numbers. That outcome, however, is hardly a ringing endorsement for the greater pacifism of the state.

Even if we leave aside problems of evidence and interpretation, there is a much more fundamental objection to numbers like the ones in Figure 2.7, and it has to do with basic statistics. Societies, whether nonstate or state, generally experience some fluctuation in combat fatalities over time, just because some element of chance is involved in their timing and magnitude. And chance can play a much larger role in small societies than it can in large ones.

To grasp the intuition behind this point, consider modern opinion surveys. Although surveys carried out on a truly random sample of a larger population will always be representative of the population as a whole, smaller surveys give answers that are considerably less precise. If we were to ask ten random Americans, for example, how much or how little they approve of the president's performance on a 0–100 scale and find that their responses averaged out to 56%, our best guess as to the president's nationwide approval rating would be 56%. That estimate could easily be off by 30% or more, though, just because the luck of the draw could easily produce ten fairly unrepresentative respondents. As we sample more and more people, however, our estimate of the president's national approval rating becomes increasingly certain. If we were to sample ninety more people, the influence of those first ten would be diluted, and we could be more certain of our answer. In fact, our margin of error would be down to roughly 10%. By the time we get to a typical commercial sample size—somewhere around 1,000 to 2,000 people—we can gauge the president's national popularity rating with surprising accuracy: most sample averages will have a margin of error of around 2 to 3%.

Just as chance can play a bigger role in survey results when the number of respondents is relatively small, chance can also play a larger role in the mortality rates of small tribes than it can in large states. Consider, for example, the most unfortunate group on the list: the Kato (or Cahto) Indians of Northern California. The Kato numbered perhaps 1,100 in the 1840s,[14] when they experienced the remarkable war-death rate of 1,500 people per 100,000, or 1.5% per year. For a group of 1,100 people, though, the loss of 1.5% of the population represents the death of about sixteen individuals in a year. That's not nothing, of course, but it's also not a terribly unlikely outcome for a society of that size, even one that's fairly peaceful. Similarly, the Goilala lost twenty-nine individuals to violent death over thirty-five years—an outcome made startling only by the fact that they numbered no more than 150 at the beginning of the period (Gat, 2008, 130). The numbers for prehistoric groups are even smaller. Ferguson dutifully notes the number of individuals upon which the percentages in Pinker's prehistoric list are based: fifty-nine individuals at Nubia site 117, twenty-three individuals at Brittany, eight at Sarai Nahar Rai, India. Even if three of those eight individuals did die as a result of warfare (and Ferguson argues that only one is unambiguous), it is not difficult to imagine chance playing a role in their demise: A statistician using a simple binomial test would tell you that three violent deaths in a sample of eight people (37.5%) is consistent with a rate of violent death in the larger population that's as low as 8.5%. If Ferguson is correct and only one of the eight deaths is unambiguously due to violence, the rate of violent death in the larger tribe could be as low as 0.3%. In short, it's very hard to rule out the possibility that these large percentages are just the result of chance variation.

To bring this point home, consider Figure 2.8, which I have created based on data from the United States Centers for Disease Control. The dark bars at the bottom represent the rate of death by poison for nine American states during the period 1999 to 2011. (Poisoning is a broad category that includes not just murder or suicide by poison but also accidental poisonings and drug overdoses.) The lighter gray bars at the top represent the rate of death by poison for twenty-seven American counties during the same period.[15] The resemblance to Figure 2.7 is remarkable. But it cannot possibly mean that residents of counties are more prone to death by poisoning than residents of states because *every*

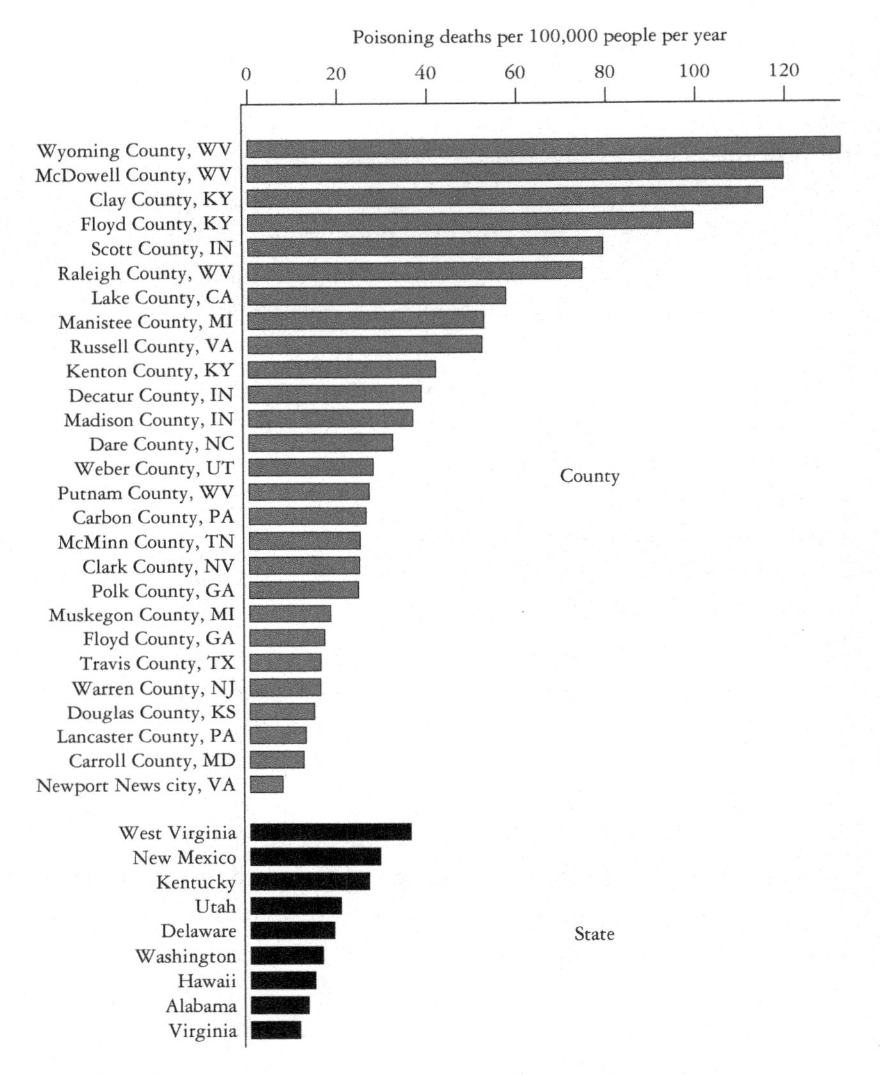

FIGURE 2.8 Deaths by poison in the United States, 1999–2011.

resident of a county is also a resident of a state. The top and bottom of the graph chart mortality rates for exactly the same people; the only difference is whether the data are aggregated by county or by state. Because counties contain far fewer people than states do, the rate of death in counties varies a lot more than the rate of death in states does. And because counties are more numerous, there are more opportunities for small counties like Wyoming County, West Virginia—a rural county characterized by deep poverty and a crushing drug problem—to register off-the-charts death rates.

It would be absurd, of course, to conclude from these data that the residents of counties are more likely to be poisoned than the residents of states. Yet numbers very much like these support the conclusion that nonstate societies are more warlike than states.

The recognition of precisely this problem led two different teams of scholars (Falk and Hildebolt, 2017; Oka et al., 2017) to explore the question of whether the apparent decrease in warlikeness of humans in state societies was simply an artifact of society size. Neither found any significant difference in conflict lethality between small-scale societies and contemporary states, once population size had been taken into account. One conclusion was that "humans from nonstates are neither more nor less violent than those from states" (Falk and Hildebolt, 2017, 805). I'll return to their findings in chapter 5, where they become important for understanding the drivers of change in one measure of the deadliness of modern warfare.

Trends in Battle Deaths

Another one of the quantities forwarded as evidence that war is in decline is the decrease, since the 1940s, of battle deaths—either in absolute terms or as a fraction of world population.[16] The best illustration of these trends comes from a graph in a 2011 *Wall Street Journal* article written by Pinker (2011b), which I've reproduced here as Figure 2.9. The graph shows a general downward trend in worldwide battle deaths per year throughout the period.

The problem here is that the number of battle deaths per year worldwide reflects a lot of different things. The four biggest ones are:

- the number of wars
- the intensity, or deadliness, of wars
- the duration of wars
- the population of the world

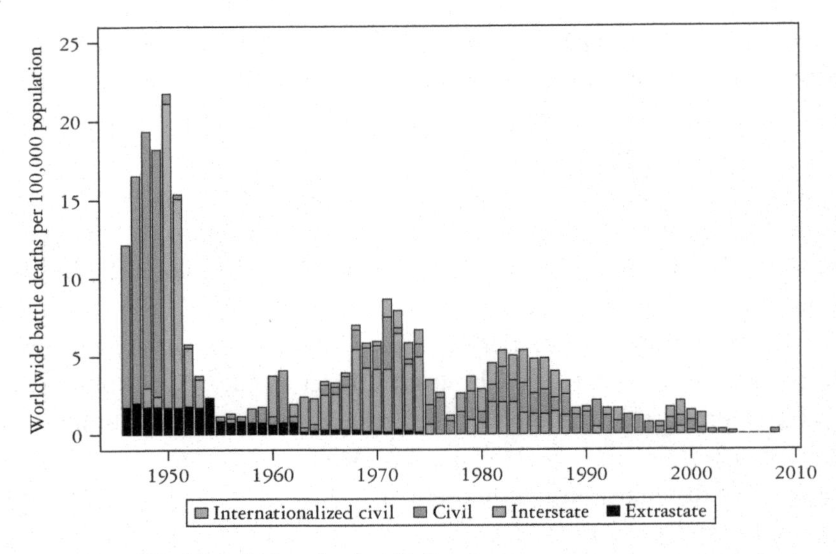

FIGURE 2.9 Worldwide battle deaths by year.

Unfortunately, we probably don't care about some of these. Even more unfortunately, we can't disentangle them using only the information in this graph.

In practical terms, what that means is the following: Imagine that we have two wars that take place a few years apart. The first war, chronologically speaking, lasts for two years and produces a half-million deaths. The second war lasts for five years and produces a million deaths. Very few observers would conclude that the world is getting *less* violent based on these two data points: the second war produced twice as many fatalities as the first! But measuring violence in terms of fatalities per year masks the intensity of the second, deadlier war and produces the illusion of a decline in violence.

In fact, precisely this pattern occurs during the time period in question. The biggest war during the early postwar period was the Korean War (1950–1953), which produced nearly a million battle deaths. If we isolate the deaths from the Korean and Vietnam wars in Figure 2.10, we can see a big spike at the onset of the Korean War, followed by a fairly rapid dropoff over the next two years. So far, so good.

Now let's consider the Vietnam War, which began in the mid-1960s and produced an even more staggering death total—something on the

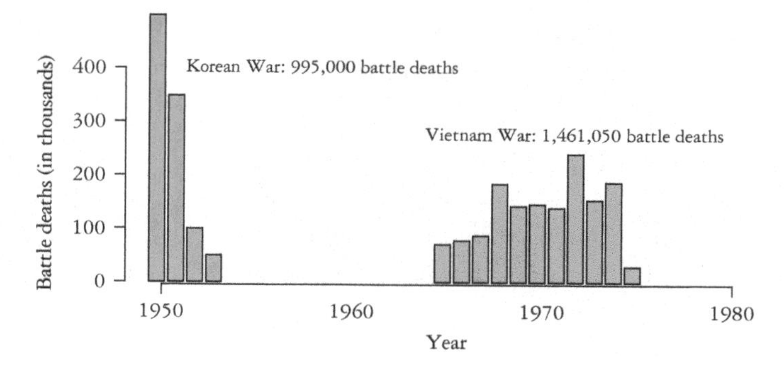

FIGURE 2.10 Battle deaths by year in the Korean War and the Vietnam War. The greater duration of the Vietnam War creates the illusion of a downward trend in war intensity, when in fact the Vietnam War produced 48% more battle deaths than the Korean War.

order of 1.5 million battle deaths. By any rational accounting, Vietnam was the more intense war. But the more modest annual death totals in Vietnam produce the illusion of a downward trend in battle deaths. That's because, relative to the Korean War, the Vietnam War produced a much steadier death toll, and it produced it over a longer period. Korea looks incredibly deadly, and Vietnam seems less so, solely because the Korean War was short and intense while the war in Vietnam was long and drawn out.

So the main point to underscore here is that looking at the number (or rate, or percentage) of battle deaths on an annual basis is a very bad idea if what we want to know is whether wars have become more or less deadly. The Vietnam War was not less deadly than the Korean War—in fact, the opposite is true. But the fact that the war in Vietnam was more drawn out produces the illusion that the world was less warlike in the late 1960s and early 1970s than it had been in the early 1950s.[17] Measuring battle deaths in a given year tells us how quickly the wars taking place in that year are killing people. It tells us nothing about how many people a specific war kills, and it certainly tells us nothing about whether that number is increasing or decreasing from one war to the next.

Moreover, even if we ignore the issue of different war durations, we still have the problem that war intensity follows a very unusual distribution that makes it difficult to eyeball data and say with any confidence that a real change has taken place. I'll get into this point in more depth below and expand on it at much greater length in chapter 6, but for the moment let me just note that there are a lot of small wars and a very small number of really, really big wars. That means that it's very hard to look at a figure like Figure 2.9 and state with any confidence that a real decline in the average has occurred—that is, that we're looking at the signal rather than the noise.

In fact, in a brief and blunt set of comments on Professor Pinker's book, Professor Nassim Nicholas Taleb (2007)—a student of probability who is best known for his popular work on the shockingly large "black swan" events that can emerge from thick-tailed distributions—offered a sketch of an argument that Pinker's data could not possibly support the conclusion that the intensity of war was decreasing over time. Professor Pinker's (2012) spirited reply prompted Professor Taleb to engage in a more formal review of the available statistical evidence. His calculations demonstrate that the apparent drop in fatalities after 1945 that we see in Figure 2.9 is, as he put it, a "natural optical illusion after a large deviation" (Taleb, 2015).[18]

The exchange between these two scholars, beyond this brief outline, is too inside-baseball to warrant much more discussion for the moment, except to note that the nontechnical parts bear closer resemblance to the sort of smack-talking that one expects prior to a professional wrestling match than they do with traditional academic discourse. The larger points, however, are worth underscoring: first, that despite popular impressions to the contrary, Professor Pinker concedes that the intensity of war was actually on the *rise* until 1945, and second, that Professor Taleb's analysis points to the conclusion that it did not decline thereafter.

The Pitfalls of Per Capita

A secondary, but no less important, point to make about this graph has to do with the use of world population as the denominator in the measure of the deadliness of war. Battle deaths as a percentage of world population is meant to capture the risk of death from war for an average human being. Similar calculations are used, for example, when we want to know whether cancer or heart disease is becoming more or less prevalent as a cause of death worldwide.

The problem is that wars and diseases are different in important ways. Unlike deaths from heart disease and cancer, deaths in war are intentionally inflicted by humans on other humans. At some point in every war, people decide to stop fighting. That decision is based, at least in part, on the human cost of the war up to that point. Accordingly, any change in decision makers' sensitivity to war costs should be reflected in the measure of the human cost of war that was actually employed by decision makers in arriving at that decision.[19] A good argument can be made that people think of war deaths either in absolute terms or, implicitly, relative to the populations of their countries. A loss of 1,500 soldiers means something very different to the Chinese than it does to the residents of the Bahamas. (In fact, it would represent the complete destruction of the armed forces of the Bahamas.) It's harder to make the case that the size of the global population is relevant to the decision to escalate or end a war.

Another difference is that, while all human beings are in principle at risk for death from cancer or heart disease, assuming they live long enough, the same is not true of war.[20] It makes very little sense to count among a war's survivors those people who, sitting at home on the other side of the globe, were comfortably beyond war's reach.

As the eminent historian Sir Lawrence Freedman (2017, xiii) put it,

> Knowing the proportion of the total world's population killed by war (and violence more generally) is unhelpful if the aim is to understand social and political processes. Numbers need to be related to particular contexts. Even during the Second World War some parts of the world were barely affected by hostilities. Governments and individuals do not assess risks by reference to global possibilities but to actual situations. To know that one is living at a time when less than one per cent should expect to die in battle is of little value when facing a heavily armed enemy any more than it is of interest for a new mother in Africa to know the life expectancy of babies in North America.

Finally, it's much harder to draw meaningful conclusions about the causes of a change in the per-capita rate of death from war than it is from rates of cancer or heart-attack deaths. If we see per-capita heart attack deaths decline, for example, we can reasonably attribute the change to improvements in some combination of diet, exercise, smoking, and medication. A change in the per-capita rate of death from war, by contrast, could happen for any number of reasons.

War and Peace

All of these reasons for skepticism raise a puzzling question. The central people in the decline-of-war literature are very smart. Their work is cited often. It might seem implausible that they could have been entirely wrong on such a straightforward question.

In fact, I suspect they're not entirely wrong. I do believe that they are mistaken when it comes to trends in warfare, and I'll explain why in Part II of this book. But I also suspect that they made the arguments that they made in part because they failed to differentiate carefully enough between war and peace.

In order to make sense out of that statement, I have to go back to the beginning of the peace science movement and the scholarship of Johan Galtung, the Norwegian sociologist and mathematician who founded the Peace Research Institute Oslo (PRIO) and the *Journal of Peace Research.* In a 1967 manuscript titled "Theories of Peace," Galtung makes an important distinction between what he calls "negative peace" and "positive peace." Negative peace is the absence of organized collective violence. This is the sort of peace that countries like North and South Korea have experienced following the ceasefire that ended their hostilities in the Korean War: they're not actively killing one another, but that's about the best you can say about the situation.

Positive peace, according to Galtung, is quite different. As he puts it: "This is peace as a synonym for all other good things in the world community, particularly cooperation and integration between human groups, with less emphasis on the absence of violence. [T]he concept would exclude major violence, but tolerate occasional violence." Galtung clearly sees positive peace as being the more profound and meaningful form of peace—a positive and cooperative relationship, rather than the simple absence of conflict.

In fact, as Professors Gary Goertz, Paul Diehl, and Alexandru Balas (2016) have shown in their recent book, *The Puzzle of Peace,* peaceful relationships in the positive-peace sense have been on the rise throughout the Cold War and into the post–Cold War period. By their measure, the relationships of 7.2% of all politically relevant states could be characterized by positive peace at the beginning of the Cold War. By the end of the Cold War, the percentage had more than doubled, to 16.6%. After the end of the Cold War, the percentage remains roughly the same, but of that 16% a majority of states transition from "warm peace" (the lowest level of positive peace, characterized among other things by the absence of war plans against one another and the presence of significant commonalities of interest) to a Deutschian security community (Deutsch et al., 1957; Adler and Barnett, 1998b), in which we see a degree of integration, coordination, and dependable expectation of peaceful change.

I think they're largely correct in that assessment, though it's hard to say much more at this point because this is a very new line of research, and the field has yet to come to anything like a consensus regarding the question of how best to measure peace.[21] At the same time, it's important to remember that positive peace, in the sense that both Galtung and Goertz, Diehl, and Balas intend it, is more than just the absence of war—in fact, Galtung himself argues that peace does not actually preclude conflict. And here we get to a crucial point: Precisely because peace is not synonymous with the absence of war, the spread of peace does not automatically imply a decline in war.

So while more peace does not necessarily imply less war, it does raise the interesting question of how they can spread at the same time. My answer to that question has to do with the logic of international order—a subject that I'll explore in considerably greater depth in chapters 8 and 9.

Conclusion

I've raised some serious objections to the evidence used in favor of the decline-of-war thesis in this chapter. If you're like me, you'll find them individually concerning and collectively compelling. In short, as much as I'd like to believe the decline-of-war thesis, the key evidence that has been advanced to support it is so deeply problematic that I can't help but be extremely skeptical.

In the same way, the arguments offered in support of the decline-of-war thesis also raise some serious concerns. I address those in the next chapter.

3

Reasons for Skepticism, Part II: Explanation

WHILE THE OBJECTIONS IN the previous chapter provide grounds for considerable skepticism on specific points, there are also three broader objections to consider regarding the explanations for a decline that have been offered. They are somewhat more amorphous than my objections to the data, but given how broadly they apply to the logic and evidence of the decline-of-war, thesis, they are in many ways even more damning.

The first objection has to do with the fact that the international system is a *complex system*. It consists of many countries, which themselves consist of people, and the interactions of those people and those countries in the context of the international system produces behavior that can't be understood or predicted just by examining people or countries in isolation. For Professors Pinker and Mueller, human beings are the heroes of the decline-of-war story: Our increasing aversion to violence has, they claim, produced a more peaceful world. As appealing as that story is, it ignores the very significant role of the social, domestic, and international context with which those people act—contexts that scholars have repeatedly shown to have a big influence on behavior. Once we incorporate those contexts into the story, it becomes much harder to believe that human pacifism, even if it is on the rise, can really play a big role in influencing war at the international level.

The second objection is that the focus in Professor Pinker's work in particular on the spread of Renaissance humanism and empathy results in a story that's a bit too neat and clean to be true. The Enlightenment gave rise to a remarkable diversity of ideas, not all of which lend themselves to peace. Only focusing on those that do runs the risk of circular reasoning—concluding that Enlightenment ideas caused peace because we only examine the ideas that are prevalent in peaceful communities. Moreover, the best-case examples for the argument that Enlightenment humanism causes peace, developed Western countries, have been shown to be very unrepresentative of the broader global community.

The third objection is that virtually none of the data analyses supporting the decline-of-war thesis takes into account the role of chance. Even if the baseline rate of something (like conflict initiations over time, or battle deaths in war) remains the same over time, there are still random fluctuations from one year to the next or from one war to the next. We need to know on average how big those random fluctuations should be if we want to know whether the baseline rate has really changed. Statisticians have provided us with an excellent and diverse array of tools to use in order to answer exactly this question, but for the most part those tools have not been brought to bear.

The Complexity of International Behavior

A complex system is a system that contains interacting agents whose behavior differs from that which one would expect solely based on their characteristics. To put it more succinctly (if less precisely), the behavior of the whole is more than just the sum of the behavior of the parts.[1]

Humans exhibit complex behavior even in very small groups. A brief story will illustrate this point. Years ago, a company started putting out motivational posters—simple photographs with thick black borders, one big word ("Teamwork," perhaps), and a one-line motivational saying underneath it, designed to be framed and put up in workplaces to inspire the people who work there. A website calling itself Despair, Inc. immediately started producing parodies of those posters, with funny lines designed to undermine motivation. If my social media feed is any indication, the parodies soon became vastly more popular than the originals.

My favorite parody poster is one that features a photo of a circle of business-attired people extending their hands inward to the center, one on top of the other. The wording at the bottom reads, "MEETINGS:

None of Us Is as Dumb as All of Us." If you've spent much of your professional life in meetings or on committees, this probably resonates with you. I certainly have, and I think of this poster whenever I find myself wondering how groups that contain such smart individuals can arrive at some of the decisions that come out of them.

That's complexity. Would any one of us have concluded that this is the best decision? Nope! Are we satisfied with it? Not entirely. Crucially, if you had asked each of us what decision we thought made sense prior to the meeting, could you have predicted the aggregate outcome? Most likely not, unless we were deciding a very simple matter.

Countries, of course, are much more complex than committees, and the international system is more complex than countries.[2] So even if it could be established that norms of peaceful behavior are spreading throughout the international system, the decline-of-war thesis makes a huge logical leap when it assumes that those attitudes will translate directly into a decrease in violence in the international system. The context of interaction matters a lot to the outcome, and there are a lot of contextual "layers" between individual human beings and international conflict that could overwhelm the impact of peaceful dispositions. I explore three of those contextual layers below—one at the level of the individual, one at the level of the country as a whole, and one at the level of the international system.

Dispositions and Situations

Most people overestimate the impact of dispositions and downplay or ignore the impact of situations in their explanations of human behavior. If we hear what sounds like a woman in an adjacent room falling down and crying out in pain, for example, our inclination is to believe that we'd get up and find out whether she needs help. More than that, we tend to believe that other people's behavior—whether or not they get up to help—is a reflection of their character. Kind people (like us!) will get up to find out whether the woman in the next room needs any assistance, but only cold or indifferent people will ignore her.

In fact, as Bibb Latané and Judith Rodin (1969) demonstrated in a landmark study, whether or not people get up to help depends crucially on situational factors. Latané and Rodin recruited male undergraduates at Columbia University to come into an office and fill out a detailed marketing questionnaire. Some filled out the survey alone; others did so in the presence of another student who was ostensibly doing the same thing but in reality was part of the experiment. While they were doing so, they were able to hear the female "market research representative" in the next room shuffling papers, opening and closing drawers, and so on. Four minutes into the experiment, they heard the representative climb up on a chair to reach for a stack of papers and then come crashing to the ground as the chair collapsed. They then heard the representative say, "Oh, my God, my foot... I... I... can't move it. Oh... my ankle. I... can't get this... thing... off me." If the students didn't intervene, they heard the representative's cries of pain gradually subside until she got up and limped out of the room. If there was another student present as part of the experiment, he did not get up to help and responded only minimally to anything that the experimental subject said or did.

Most of the students who were left to fill out the survey alone—70%—got up and attempted to help the woman in the next room. When there was another student there and that student failed to react, however, that behavior changed drastically: only 7% of the subjects got up to help. A relatively minor change in the situation—the existence of a bystander who did nothing—was sufficient to produce a night-and-day difference in behavior.

Our human tendency to attribute behavior in such situations to the character of the individual in question rather than to the specifics of the situation is so profound that psychologists refer to it as the "Fundamental Attribution Error."[3] That it is an error is no longer in much doubt:

> Consider the following scenario: While walking briskly to a meeting some distance across a college campus, John comes across a man slumped in a doorway, asking him for help. Will John help him, or will he continue on his way? Before answering such a question, most people would want to know more about John. ... In fact, however, nothing one is likely to know or learn about John would be of much use in helping predict John's behavior in the situation we've just described. ... A half century of research has taught us that in this situation, and in most other novel situations, one cannot predict with

any accuracy how particular people will respond. At least one cannot do so using information about an individual's personal dispositions or even about that individual's past behavior. (Ross and Nisbett, 1991, 2).

American isolationism provides an excellent illustration of the way in which the fundamental attribution error can creep into scholars' explanations of a country's foreign policy behavior. The United States between World War I and World War II is often said to have pursued an isolationist foreign policy, due in large part to the noninterventionist "mood" of its citizens. Much of the historical scholarship on isolationism in that period and subsequently has focused on figuring out which kinds of people—the less educated? The poor? Recent immigrants? Republicans? People from the Midwest?—are more likely to express isolationist sentiments.[4]

More recent research has shown, however, that this apparent disposition toward nonintervention depends crucially on the situational context. In the case of the interwar period, Americans' inclinations not to become involved in the growing European war hinged crucially on a key situational factor: Very few Americans, even those at high levels, believed that Germany constituted a real threat to the Continental powers. The shockingly successful German invasion of France in May and June of 1940 proved conclusively that Germany was a genuine threat, and as a result American support for intervention skyrocketed. While only 20–30% of Americans were willing to risk war in order to aid England and France prior to the invasion, support for doing so had cleared 70% by March 1941.[5] Dispositional explanations for isolationism cannot explain this change of heart: it's extremely unlikely, of course, that nearly half of the American population moved to the coasts, switched political parties, or went to college in the nine months following the fall of France.

It seems plausible, at least, that nonviolence is every bit as dependent on situation as isolationism is. Indeed, that is the premise of political theorist Hannah Arendt's brilliant and controversial book *Eichmann in Jerusalem: A Report on the Banality of Evil.* Arendt's horrifying conclusion was that the subject of her book, Nazi SS-*Obersturmbannführer* Adolf Eichmann, far from being incomprehensibly evil, was driven to commit atrocities by the circumstances of the time. While Arendt stopped short of endorsing Eichmann's own explanation—that his orders compelled him to comply—she concluded that he was an ideologue who was easily swept up in the Nazi movement and able to justify actions that he otherwise would have found immoral because they furthered the movement. He was, in her words, a "joiner." She writes that "May 8, 1945, the official date of Germany's defeat, was significant for him mainly because it then dawned upon him that thenceforward he would have to live without being a member of something or other."[6] Far from being unusual, Eichmann's susceptibility to influence struck Arendt as being terrifyingly common.

Stanley Milgram, a psychologist at Yale, was struck by Arendt's conclusions and set up a series of experiments to explore the limits of human obedience.[7] The subjects of Milgram's experiments were told that a person in another room—supposedly another subject, but in reality a confederate—would answer a series of questions given to him by the subject. The actual subjects of the experiment, the "teachers," were asked to push a button whenever the "learner" in the next room gave an incorrect answer. Pushing the button would administer an electric shock. The subjects were told to use a dial to make each electric shock stronger than the previous one, up to a maximum of 450 volts. The shocks, according to the experimenter, would cause no lasting damage. (This, at least, was true: the shocks were fictional. The button activated a tape recorder that had been loaded with a pre-recorded script.) At first, pushing the button produced the sort of sounds from the next room that one might expect from someone who had just been zapped unexpectedly with a minor shock. Before long, however, the subject could hear the person in the next room banging on the wall and demanding that the experiment be brought to an end. If the shocks continued, the reactions from the next room became more intense and the subject could hear pleading, as well as the mention of a heart condition. Beyond a certain point, pushing the button produced no sound from the next room at all. If the subject expressed discomfort at the noises coming from the other side of the wall (as each did, sooner or later), the experimenter simply urged the subject to continue.

Milgram and his colleagues expected that very few subjects would administer apparently murderous voltages of electricity to another human being simply because they were told to do so. In fact, 65% of the participants did. They argued with the experimenter, who remained bizarrely impassive in the face of the cries from the next room. Some got up as if to leave but dutifully returned to their seats when told to do so. But in the end, 65% of the subjects continued to increase the voltage and push the button until no more sounds came from the next room.[8]

There is substantial evidence that context drives behavior outside of the laboratory as well. Author Karl Marlantes is a decorated veteran of the Marine Corps who served in the Vietnam War. His account of his time there, published under the title *What It Is Like to Go to War* (2011), is an extended meditation on the effects that war had on him.

Marlantes offers an account of an airstrike that is particularly relevant. In it, he helped protect a team of Marines that had taken a hill and were trying to defend it against their North Vietnamese pursuers until a helicopter could arrive to pick up one of their gravely wounded comrades. Marlantes called in air support in the form of some Marine A-4s, which came loaded with bombs and napalm. They used both, and the North Vietnamese soldiers were left scattered across the hill, charred from the napalm, either dead or dying.

At the time, Marlantes writes, he was elated at having saved his team. Looking back, he writes, "I now think of what was 'the enemy' as human beings, so I find it hard to crow about burning them to death." He makes a point of noting that he now feels an empathy for his enemy that he did not feel at the time—the same empathy that, according to Pinker, has grown along with the spread of Renaissance humanism.[9]

The key question is, what differences does that empathy make? Looking back on his experience at the time, with the benefit of years and the empathy that they provide, would Marlantes do anything differently today? By his own account, he would not:

> I'd still do the same thing, only I would be aware of a horrible dilemma. I would be much more reluctant to use napalm now, knowing I could get the job done a lot more humanely with bombs. But scrambled aircraft arrive on station loaded with what they're loaded with. Once I had decided to be in that situation, I couldn't then decide that the team should sacrifice itself for my misgivings about using napalm. ... Ideally, I would hope that, in spite of the adrenaline, I'd at least stay conscious of a terrible sadness while I

> burned these people. But burn them I would. (Marlantes, 2011, 41–42).

To his credit, Marlantes is ruthlessly honest. Looking back, he does feel empathy for the other human beings on the hill. He no longer sees them as the impersonal, inhuman enemy. But these other human beings are doing their best to kill *his* human beings, the soldiers on his side of the war. And in that situation, while empathy might produce a lingering sadness, it would not prevent him from using napalm to kill the North Vietnamese soldiers. The requirements of the situation, in other words, outweigh the dispositional compulsions of empathy.

It might seem that I've chosen an easy case to use to demonstrate the power of situations over dispositions because war is an unusually compelling environment. That's probably true. But it's also the environment that's most relevant to the decline-of-war argument. Ignoring the power of the situation is a bad idea in general when we're trying to explain human behavior, but it's *really* a bad idea when we're trying to explain how people behave when the lives of their comrades or their fellow citizens are on the line.

So, the first point to make (and I hope I have not made it in tedious detail, but I do think that it bears strong emphasis) is that dispositions alone rarely predict behavior. Even a soldier looking back at his actions in Vietnam with sadness and empathy recognizes the fact that he would have to do precisely the same thing today if he were to find himself in the same situation. A majority of ordinary Americans administered what they thought were lethal doses of electricity to their fellow human beings when someone in a position of authority asked them to do so. We don't like to think that we, ourselves, are so malleable as to be capable of these sorts of behavior, but decades of research point unambiguously to the conclusion that we are.

Strategic Interaction

Even if we entirely leave aside the power of the situation in determining human actions and assume, *contra* lots and lots of evidence, that individual dispositions translate unproblematically into actions, we immediately run into another problem: People and countries react strategically to one another's actions. A change in the behavior of one actor may result in a change in the behavior of others, and the outcome of those changes may be the opposite of what we'd expect. For that reason, there's no guarantee that more peaceful preferences in the context of strategic interaction will actually lead to more peace. They may actually do the opposite.

One of the most well-known examples of this outcome is the onset of the First World War. When Austrian Archduke Franz Ferdinand was assassinated and the Austro-Hungarian Empire contemplated war against Serbia, Germany had to balance the merits of backing an ally in what they hoped would be a short Austro-Serbian war in the Balkans against the risk of a broader Continental war. The German decision to back Austria was based in large part on the mistaken belief that the other powers would not intervene: As the Saxon envoy in Berlin put it, "England is absolutely pacific and France as well as Russia likewise do not feel inclined towards war" (Clark, 2013, 515). Perversely, if the British or the Russians had been more aggressively interventionist early on in the crisis, Germany would almost certainly have tempered its support for Austria-Hungary, and the Continent might have remained at peace.[10]

Of course, one could also cite plenty of examples of cases in which a bellicose foreign policy also got a country into war: I don't mean to imply that an increase in pacifism *inevitably* increases the risk of conflict! Rather, when we take the context of strategic interaction into account, we can't simply draw a direct connection between an increase in pacifism, on the one hand, and less conflict on the other. We have to take the calculations of other actors into account, and those calculations could easily lead to more aggressive actions that would increase the probability of war.

Professors Andrew T. Little and Thomas Zeitzoff lay out a detailed theoretical model that nicely captures the logic of how countries' preferences and calculations can coevolve when war becomes more or less costly and countries become more or less pacifistic as a result. The logic is a bit more complex than what I've laid out above, but the upshot is similar: as countries become less willing to fight, they do more *actual fighting*. The authors argue specifically that the logic of their argument undermines the decline-of-war thesis: "[E]ven if we accept that conflict is declining over time as a result of an increasing relative value of cooperation, we can not infer from this fact that people have become more peace loving" (Little and Zeitzoff, 2017, 17).

Structure and Anarchy

As important as the immediate situation and the logic of interaction can be in determining an individual's behavior, there are other contexts that matter as well. To international relations theorists, perhaps the most important of these is international anarchy. The word "anarchy," in this context, does not connote chaos or disorder; rather, it refers to the lack of any overarching political authority in the international system. While an experimenter's office or a war zone may be the context within which people interact, anarchy is the context within which countries interact.

The importance of anarchy lies in the incentives that it produces for the countries that make up the international system. In the absence of a global police force capable of righting wrongs as serious as invasion and conquest, no country can be assured of its own survival. Each must prepare, to some degree, for the possibility of conflict with others, no matter how peaceful its intentions. Because those intentions can never be known with certainty and countries concerned about their own survival are often risk averse, that preparation in and of itself can be seen as threatening. This situation, in which actions taken to increase the security of one country prompt reactions that increase tensions and create conflict, is known in the international relations literature as the "security dilemma."[11]

Without a doubt, the classic modern work on the impact of anarchy on international politics remains Kenneth Waltz's *Theory of International Politics*. While Waltz did not deny that human nature and domestic politics can play a role in determining a country's foreign policy,[12] he argued that they were typically swamped by the logic of survival under anarchy.

> The state among states, it is often said, conducts its affairs in the brooding shadow of violence. Because some states may at any time use force, all states must be prepared to do so—or live at the mercy of their militarily more vigorous neighbors. Among states, the state of nature is a state of war. This is meant not in the sense that war constantly occurs but in the sense that, with each state deciding for

> itself whether or not to use force, war may at any time break out. . . . Among men as among states, anarchy, or the absence of government, is associated with the occurrence of violence. (Waltz, 1979, 102).

The idea that anarchy stood in the way of a reduction in international warfare was also endorsed by none other than Professor Norbert Elias, the German sociologist whose work on the "civilizing process" is one of the biggest influences on Professor Pinker's book:

> At the international level there is no overarching power to prevent a stronger state from invading a weaker state to demand taxes and obedience from its citizens and so de facto to annex the weaker state. Nobody can prevent a mighty state from doing this except another mighty state. And if such states exist they live in constant fear of each other, in the fear that their rivals could become stronger than themselves.[13]

Waltz's book prompted countless counterarguments, modifications, clarifications, extensions, and heated debates. Robert Axelrod (1984) argued that cooperation among self-interested actors could easily evolve despite anarchy, thanks to the promise of future interactions—the so-called "shadow of the future." Robert Keohane (1984) argued that an evolving web of international institutions was increasingly able to facilitate international cooperation. Later, Alexander Wendt (1992) took a considerably more radical position, arguing that, as the title of the article puts it, "anarchy is what states make of it": anarchy is not an inherent feature of the international system but is rather a social construction that is amenable to reinterpretation.

It is no exaggeration to say that these and related works laid the foundation for international relations theory for decades. A summary of the research that came in their wake would take multiple chapters. The main point, though, is that much of the debate centered on the impact of anarchy on the behavior of the countries that make up the international system and how that impact might be mitigated.

At a minimum, the logic of international anarchy calls into question the claim that a nonviolent public will automatically produce a nonviolent foreign policy. As the old Latin adage goes, *Si vis pacem, para bellum* (If you want peace, prepare for war). Even strong advocates of nonviolence can find themselves compelled to advocate military means to defend the national interest, and the defense of the national interest is a constant concern in the absence of world government. Self-defense can easily be mistaken for aggressive intent and produce precisely the conflict that it is meant to deter. For the most part, Pinker and Mueller, the main advocates of the norms-of-nonviolence explanation, simply don't address international anarchy or explain how the peaceful dispositions of the citizens of a country could prevail in an international context that compels countries to risk conflict.[14]

The Spread of Enlightenment Humanism

Another reason to be skeptical of the argument that underpins the more ambitious variant of the decline-of-war thesis is that the causal story that drives it—the spread of Renaissance humanism and empathy—is a bit too neat and clean to be true. The idea that human progress marches steadily in a single direction, albeit with occasional slight stumbles and reversals, is an example of what the eminent British historian Sir Herbert Butterfield (1965, v) called "the whig interpretation of history":

> the tendency in many historians to write on the side of Protestants and Whigs, to praise revolutions provided they have been successful, to emphasise certain principles of progress in the past and to produce a story which is the ratification if not the glorification of the present.

Regardless of the debate, Butterfield (1965, 5) writes, "The historian tends in the first place to adopt the whig or Protestant view of the subject, and very quickly busies himself with dividing the world into the friends and enemies of progress."

Whiggish accounts of history suffer from multiple shortcomings. In the first place, they tend to view the past through the lens of the present and overemphasize those elements of history that are comprehensible and sympathetic to their authors. For that reason, the understanding of the past that they convey is biased in favor of the ideas and institutions of the present. That bias tends to produce a narrative that emphasizes historical progress toward those ideas and institutions rather than the usually more complex, usually more alien understanding of history as it actually happened.

Such accounts are also teleological, meaning that they tend to portray history as progressing toward a particular goal rather than as being driven toward an uncertain future by the ideas, resources, and passions of the peoples who comprise it. Hindsight is always 20/20, as the saying goes, but Whiggish hindsight is even more acute and considerably more narrow. Historians generally struggle against this artificial clarity, and with good reason.

Professor Pinker is clearly aware both of the charge of Whiggery and of its implications. He (2011*a*, 692) writes:

> The metaphor of an escalator, with its implication of directionality superimposed on the random walk of ideological fashion, may seem Whiggish and presentist and historically naïve. Yet it is a kind of Whig history that is supported by the facts.

Much of this book is devoted to an evaluation of that last claim, so I'll leave it to the reader to decide whether or not it's warranted. I will note in passing that it seems to me to be an odd defense: Few if any Whig historians *don't* believe their accounts to be supported by the facts.

While the charge of Whiggery is a fairly abstract (one might even say "academic") objection, it has real consequences. I briefly explore three below: an unwillingness to recognize those aspects of Enlightenment thinking that aren't especially pacifistic; the tension between perpetual peace and just war; and the fact that, even now, Enlightenment values are limited to a fairly small and unrepresentative part of the globe.

The Dark Side of the Enlightenment

The Enlightenment was an eighteenth-century movement that gave rise to some of the most powerful and enduring ideas of the modern era, many of which—representative governance, freedom, progress, an emphasis on reason, tolerance—are so fundamental to liberal democracy that many citizens simply take them for granted. It would be absurd to argue that the Enlightenment did not represent a substantial leap forward in human progress or that much of the betterment of the human condition that has taken place over the past two centuries does not owe a very considerable debt to the ideas that came out of the Enlightenment.

That said, the Enlightenment gave rise to a tremendous diversity of ideas, not all of which turned out to be either liberal or especially benign. The philosopher John Gray (2015), in his critique of *Better Angels,* makes this point more expertly than I will here, but the upshot of my complaint is this: The Enlightenment produced a big, complex, often contradictory body of ideas that don't all lead to outcomes that are liberal, peaceful, or progressive.

Exhibit A for this argument has to be Jean-Jacques Rousseau. Rousseau, one of the best-known of all Enlightenment thinkers, was also one of the most articulate critics of progress. He is most prominently associated with the argument that modernism, science, and progress are corrosive to morality. Johann Gottfried Herder, undeniably an Enlightenment philosopher, is perhaps the most central philosopher of nationalism, an idea that cuts directly against the universalism that is typically associated with Enlightenment thought and that is rarely associated with the spread of peace. Georg Wilhelm Friedrich Hegel's emphasis on freedom derived from the writings of such philosophers as Rousseau and Immanuel Kant, but in *Elements of the Philosophy of Right* he argued that true freedom can only be achieved via a strong state—an idea that would be anathema to modern libertarians but was welcomed by twentieth-century fascists as well as by Karl Marx, who adapted Hegelian ideas about historical progress to produce an entirely different brand of totalitarianism. If nothing else, the Cold War should have dispelled any notion that Enlightenment values will save us from conflict: Marx was nothing if not a child of the Enlightenment, yet his disciples clashed, dangerously and often and at staggering cost, with those of Locke and Kant.

The argument that Enlightenment humanism leads directly to peace also ignores the distinctly illiberal ideas that arise as a reaction to the shortcomings of Enlightenment thought and against the modernity that it ushered in. This point is the central theme of Pankaj Mishra's 2017 book *Age of Anger,* which argues that much of the violence of the past two centuries has its roots in the anger of the people left behind by modernity and the philosophies and demagogues to which they were drawn. Similarly, Mark Lilla's *The Shipwrecked Mind* is an extended meditation on the power of political nostalgia and the reactionary ideas to which it gives rise, while Jan-Werner Müller's *What Is Populism?* highlights the antipluralist undercurrents that both arise from and erode liberal democracy. The bottom line in each case is that, even in democratic, free-trading societies that reflect the ideals of Enlightenment humanism, some people end up a lot happier than others, and disaffection can ripen into conflict.

Why do these examples matter? I'm not arguing that Kant didn't write *Perpetual Peace* or that the values of liberal democracy owe nothing to the Enlightenment. But when Professor Pinker argues that "liberalism, modernity, cosmopolitanism, the open society, and Enlightenment values always have to push against our innate tribalism, authoritarianism, and thirst for vengence" (Edsell, 2017), he misses the point that much of that tribalism and authoritarianism owe their present-day expression either directly or indirectly to the Enlightenment as well. That, in turn, raises the very real possibility that the relationship that Pinker and others see between Enlightenment values and peace is either circular or wrong. If the only Enlightenment ideas that are acknowledged as such are those that haven't given rise to war, the proposition that Enlightenment values cause peace becomes true by definition. If we look at the full range of ideas that came out of the Enlightenment as well as those that have arisen as a reaction against the more liberal ones, it becomes hard to sustain the claim that Enlightenment values inevitably lead to peace.

Indeed, scholars who have taken a broader and more nuanced view of the Enlightenment have often found it to be a mixed blessing at best when it comes to peace. Some even conclude that it has done more harm than good. Roger Osborne, after a sweeping review of Western civilization in which he chronicles ideologies, extremism, genocide, war, and a shocking range of examples of what Robert Burns called "man's inhumanity to man," offers a forceful and not atypical summary:

> There remains a belief, particularly among liberal westerners, that [the present moment] is simply a short-term crisis brought on by the hypocritical piety of certain leaders. There is even an idea that the current situation has been brought about by irrational, religious-based ideas, and that a healthy dose of rationalism will put us back on course. The history of the last 2,500 years, and the last 150 years in

particular, shows that this is an illusion. The fundamental western belief that there are rational ways of organizing the world which will bring benefit to all has been at the root of every human-made catastrophe that has overtaken us. (Osborne, 2006, 491–492).

I'm not a scholar of the Enlightenment by any means. But I'm pretty sure that a discussion of the Enlightenment that only covers the parts about science, reason, progress, and peace does not do the subject justice. Pinker's disagreement with a broad swath of Enlightenment scholars on the nature and implications of the Enlightenment is evident in *Enlightenment Now,* which offers the bizarre spectacle of the Johnstone Family Professor in the Department of Psychology at Harvard University taking a populist position against the intellectuals who disagree with him. Perhaps predictably, the book was savaged by precisely the Enlightenment scholars who would normally be expected to cheer him on.[15]

Just War and Perpetual Peace

It's difficult to imagine any ideology of progress leading to the abolition of war, simply because wars are often fought to further somebody's notion of human progress. Even people who subscribe to Enlightenment ideals believe in the concept of a just war, or a war that is considered to be morally justifiable.[16] For that reason, the moral underpinnings of Western liberalism can make people *more* willing to fight under some circumstances than they otherwise would be. As Rory McCann's character, Sandor Clegane, put it in *Game of Thrones*, "Lots of horrible shit in this world gets done for something larger than ourselves."[17]

One category of liberal just war is humanitarian intervention, or intervention to protect the human rights of people under threat. As Professor Gary Bass (2008) demonstrates in his engaging study of the subject, the spread of liberalism in Europe corresponded with the rise of humanitarian intervention in practice, starting with the Greek civil war in the 1820s. Recent examples include NATO's involvement in Kosovo in 1999 and Libya in 2011, as well as, arguably, the present conflict with Islamic State forces.

While some commentators do see a trend away from American humanitarian intervention starting with the George W. Bush administration (Kim 2003), global trends are moving in precisely the opposite direction. In 2005, the member states of the United Nations ratified a new doctrine known as the Responsibility to Protect (R2P). While in many ways R2P establishes a formal framework for humanitarian intervention that relies as much as possible on nonviolent means, it also establishes as a fundamental principle of international law not just the right of the international community to violate the sovereignty of independent states in order to protect their populations from human rights abuses but their actual *responsibility* to do so. Given many states' present human rights practices, if taken seriously this principle could lead to widespread intervention that would not have been countenanced or even contemplated in the absence of liberal Enlightenment norms and values. Professor Page Fortna (2013, 569) makes this point succinctly in her review of Professor Goldstein's book:

> I support R2P on ethical grounds, and think it will lead to a more just world, but there is a tension here that goes unremarked. Another word for *military intervention,* even if its motive is humanitarian, is *war.* R2P may well increase violence rather than reduce it.

R2P goes beyond intervention, of course. If intervention fails to address the root causes of human rights violations or, worse, destabilizes a state and makes future violations more likely, it will have accomplished little. NATO's intervention in Libya, for example, while hailed as a "model intervention" at the time (Fortna, 2013, 569), left a greatly destabilized country where what little security there was was provided by rival militias.

Accordingly, as the International Coalition for the Responsibility to Protect puts it, "advocates around the world have embraced [R2P] as a full spectrum of responsibility: from the responsibility to prevent, to react, and to rebuild."[18] When, as is often the case, the government of the country in question is responsible for human rights violations, measures up to and including regime change and nation building, *à la* Iraq and Afghanistan, plausibly fall under the umbrella of peacebuilding.[19]

Again, these are not necessarily bad goals. Muammar Gaddafi was massacring his people with breathtaking brutality in order to avoid the fate experienced by other leaders during the Arab Spring, and stopping him was the right thing to do. We cannot pretend, however, that stopping him did not amount to military intervention, or that the price of stopping him itself isn't still being paid in blood. A good cause may justify the use of violence, but it doesn't erase the fact of it.

The Study of WEIRD People

By far the best examples of a spread of peace in the decline-of-war literature come from the Western industrialized countries in the post–World War II era. One of Pinker's (2011*a*, 249–251) most memorable passages has to do with the number zero: No country has used nuclear weapons on another since 1945, no Great Powers have fought one another since 1953, no interstate wars have been fought by European countries, no countries have conquered parts of other countries by force (those last two have to be updated in light of Russia's 2014 war with Ukraine), and so on. Most of these zeros involve the advanced industrialized countries that Pinker argues are at the vanguard of the spread of Renaissance humanism.

This narrow focus omits a lot of people, and there's no reason to believe that the people it includes are representative of the rest. In fact, there are very good reasons to believe that they're not. In 2010, psychologists Joseph Henrich, Steven J. Heine, and Ara Norenzayan (2010, 61) wrote what has since become a very widely cited paper on exactly this issue. Titled "The Weirdest People in the World?," the article pointed out that behavioral scientists regularly draw conclusions from studies of people who are WEIRD—Western, Educated, Industrialized, Rich, and Democratic. WEIRD people, the authors argue, are "among the least representative populations one could find for generalizing about humans" when it comes to traits like cooperation, fairness, and moral reasoning—precisely the sorts of traits that Pinker argues underpin the decline of war.

As you'll see in detail later in the book, I don't think that the claims of a recent decline of war among WEIRD people are entirely wrong. But WEIRD people aren't the only ones who have managed to put a damper on international conflict: other "islands of peace" have existed at different points in time, and many of them can't be attributed to WEIRDness.

The Role of Chance

My final reason to be skeptical of the major arguments and findings that support the decline-of-war thesis is more abstract but no less important than those that I've listed so far: They generally don't take into account the role of chance.

We don't tend to think about the role of chance much when we're contemplating international conflict. We tend to assume that because things happened a certain way, they *had* to happen that way. We see history as a concrete series of events, and if a particular kind of event like the use of force between two countries happened more during one part of history than it did during another, that's all we need to know. That's the only way it could have happened.

A more sophisticated variant of this argument is grounded in the distinction between populations and samples. The wars that we're interested in are bounded by concrete dates. That means that we have, not a sample of wars, but rather the complete population of wars that occurred during those historical periods. If we have the complete population, one might argue, why are we doing statistical inference at all?

An analogy helps to clarify this position: Survey researchers typically interview only a tiny fraction of a given population, so their conclusions about the population as a whole ("Smith's job approval rating is at 32%, ± 3%") include some degree of uncertainty. But what if they interviewed the entire population? They'd be able to say that Smith's job approval rating is exactly 33.12% (or whatever). There wouldn't be any uncertainty at all about the population's views. Analogously, if you think of these sets of wars as the complete population of wars that happened during those two time periods, then it makes no sense to talk about uncertainty.

This reasoning doesn't command a majority position among the stats-and-war crowd. In fact, I can't think of any practitioner who has actually espoused it. I think that's because the overwhelming majority of us believe, at least implicitly, that the history that we have observed was not fated to happen the way that it did—that chance events could very easily have changed the outcome. To return to the survey research example, people's answers to survey questions about things like approval ratings are rarely so concrete that you'd get exactly the same answer under different circumstances. If you ask me about President Smith on the first sunny day of spring, when I'm out of my office and enjoying the weather, I might tell you that Smith is doing a bang-up job. If you ask me the same question in the dead of winter, before I've had my morning coffee, I might give you a much less charitable assessment. If that's the case more generally, even a survey of the entire population would produce results that contain some uncertainty because, if you did it over again, you'd get a different number.

So it is with war. In any war, there are critical junctures at which things might have gone a different way. Hitler's Germany might have been stopped cold by the French in 1940 and slowly rolled back to Berlin. Or, as fans of *The Man in the High Castle* would point out, things could have gone another way: As that alternative history goes, Giuseppe Zangara could have succeeded in his 1933 attempt to assassinate Franklin Delano Roosevelt, thereby eliminating both the New Deal and the Lend-Lease Act and undermining the Manhattan Project to such an extent that the Nazis develop atomic weapons first, bomb Washington, and end up dividing North America with the Japanese Empire. One of Paraguayan President Francisco Solano López's military commanders could have talked him out of his doomed war against Argentina, Brazil, and Uruguay in the 1860s, or talked him into surrender once it was clear that the war was going very poorly. Or, as I pointed out back on page 26, the Cuban Missile Crisis could very easily have escalated to nuclear war if one of the three Soviet officers on a Foxtrot-class submarine had changed his mind. When you add up all of those little chance events, it's hard not to conclude that history could very easily have played out differently—that some wars could have been far more deadly and others far less so.

As these examples suggest, chance plays a prominent role in warfare in at least two ways. First, while we can spot conflict-prone situations or regions, we cannot with any certainty predict when conflict will actually break out until it's about to happen, and sometimes not even then. The rate of conflict onset should be higher in conflict-prone situations, but that's a statistical generalization, not a point prediction.

The second way in which chance plays a role in warfare has to do with escalation. Despite the fact that war has been dissected and analyzed in hundreds if not thousands of scientific books and articles over the past few decades, we simply don't have any meaningful idea of how deadly a war is going to be before it starts. The opposing parties can't really have a clear idea about how much blood has to be shed before a negotiated settlement becomes possible. Worse, although many wars remain fairly small, a very small number of wars become shockingly large—and we don't know with any degree of certainty which wars will fizzle out and which ones will escalate to cataclysmic proportions.

This fact raises a real problem for analysts: How do we distinguish the signal from the noise? Looking only at a series of events and non-events, how can we tell whether the fundamental relationships that govern the thing that we're interested in (the amount of underlying tension that produces conflict, or the escalatory dynamics that turn a few small wars into really big ones) have changed, or whether we're just looking at the kind of chance variation that we'd expect to see over time?

This problem is exacerbated by one of those odd quirks in human psychology: We try pretty hard to spot patterns in random data. There is even a name for this predilection: *apophenia,* the human tendency to perceive meaning in random or meaningless information. The constellations are an obvious example of people's ability to pick out meaningful patterns from a random array of stars. The persistence of horoscopes, too, is a testament to our capacity for seeing patterns in our daily lives where (sorry) none exist. The number of people seeing images of Jesus in everything from toast to marmite to sliced potatoes is so well known that it has inspired a "grilled cheezus" sandwich press.

In short, we humans simply can't be trusted to eyeball something and decide whether or not it's the result of chance. We need statistical inference to keep us honest. Statistical inference is designed with exactly this problem in mind. Its value lies in its ability to estimate, very precisely, how uncertain we are about our estimate of a number, whether that number is the rate of conflict initiation or the propensity of conflicts to escalate. Once we know how uncertain our estimates are, we can answer the central inferential question in the decline-of-war literature: How can we reliably differentiate changes in violent behavior from random noise?

Randomness in Thick-Tailed Distributions

Perceiving patterns in randomness is an especially acute problem when we're dealing with thick-tailed distributions, like the distribution of battle deaths in war. Thick-tailed distributions have the characteristic that the overwhelming majority of observations are relatively small in magnitude and the small number of very large ones are extraordinarily large. Because smaller events are so incredibly common, they're very likely to trigger our apophenia and make us conclude that there's a pattern there when there really isn't. Because larger events are so spectacularly large, moreover, they can make us think that we're looking at a new pattern when we really aren't.

Consider, for example, household income in the United States, which is measured by the Census Bureau's ongoing American Community Survey (ACS). Figure 3.1 shows a sample of household incomes from the ACS's 2009–2013 Public Use Microdata Sample (PUMS), an anonymized subset of the actual survey that is made available for researchers. This particular figure shows the incomes of 1.7 million of America's 123 million households. This is a notoriously thick-tailed distribution: while the majority of the households report modest or even negative income, the top 1% earn upward of $430,000 per year.

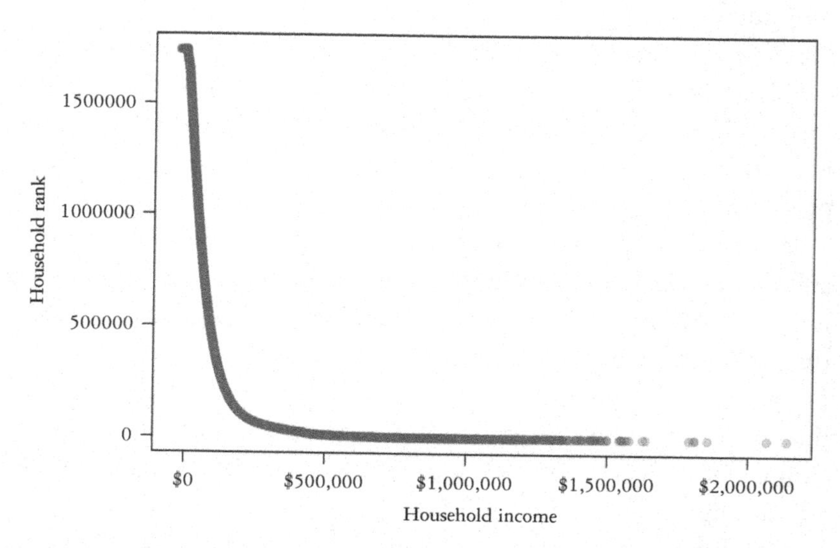

FIGURE 3.1 The distribution of a sample of American household incomes.

Even this sample understates the length of the tail in the distribution of income: if the long tail on the right captured the annual household income of software magnate Bill Gates, whose net worth increases by about $5 billion every year,[20] the rescaling of the graph would make it impossible to distinguish anyone else's income from zero.[21]

Now, let's imagine that you've been hired by the Census Bureau to contact people and ask them questions about their lives, their families, their households, and so on as part of the ACS. You manage to reach fifty respondents over the course of your first day, and you dutifully record all of their answers. At the end of the day, it occurs to you that the calls you've been making didn't actually *seem* very random: your respondents got less wealthy as the day went on. A quick plot of income vs. respondent number for your fifty respondents (Figure 3.2) confirms your impression.[22]

The first family you reached had a household income of about $115,000—higher than average, to be sure, but not shocking. The next family, though, had a household income of over $850,000, which ended up being the highest income of the day. The third and fourth households were unremarkable, but household five had an income of over $190,000 and household six brings in over $525,000! Tallying it up, you find that the first six households you reached had almost 40% of

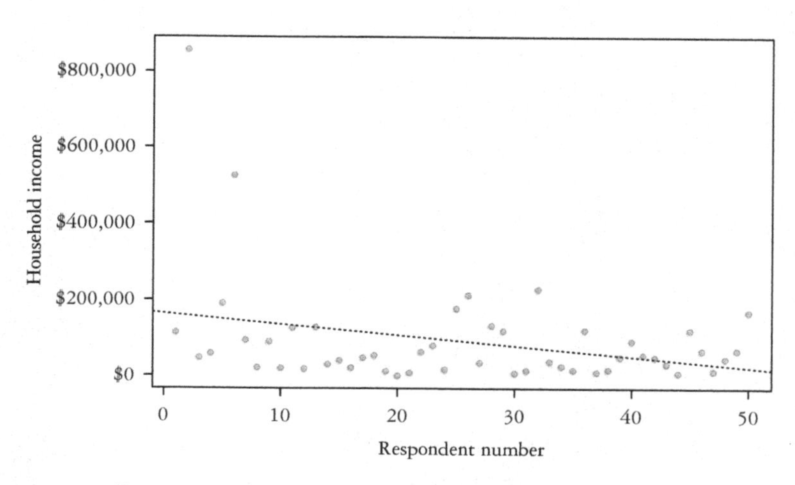

FIGURE 3.2 Illusory decline in average household income over the course of fifty interviews.

the total income of all fifty combined. That seems incredibly unlikely if this really is a random sample—and if it's not, there might be reason to worry about the entire survey. Should you contact your supervisor and explain your concerns?

The answer, as you've almost certainly surmised, is "no." It's not actually too unlikely that you'd get a couple of very wealthy respondents early on. They're rare, to be sure, but they're not *that* rare. What's more important is the fact that the distribution of income is really skewed: the largest incomes are a *lot* larger than the smaller ones. So when you do encounter a few households with really high incomes, your expectations for the rest of your calls are really out of whack. As the day progresses, you don't see any more really high incomes, and you start to wonder whether something has changed. It's normal to think that it has—the downward trend in the incomes of your respondents is pretty noticeable. But it really is an illusion: the fifty observations in Figure 3.2 were drawn totally at random from the PUMS data. It just looks a lot like a trend because the big outliers are so much bigger than we intuitively think they should be by chance—and because we humans are cursed with apophenia.

Now, with all that in mind, take another look at Figure 3.2. Once you've done so, flip back to Figure 2.9, the graph of annual battle deaths divided by world population. Keeping in mind the fact that battle deaths, like household income, follow a thick-tailed distribution, are you *really* confident that what you're seeing is a downward trend in the data and not just random noise? I'm not.

Another way to see this point is to take a look at Figure 3.3, which shows what happens when you draw random numbers from a normal distribution and a particular thick-tailed distribution called a *power law distribution* (which, not coincidentally, provides a pretty good fit to the data on the deadliness of war, as we'll see in chapter 5). The value on the y-axis is the *running mean* of these series of numbers: At observation 200, the running mean is the average of the first two hundred observations. The dashed line represents the average of the distribution from which these observations are drawn. If you want to use a sample value as an estimate of that average, the running mean should converge to the underlying average value pretty quickly.

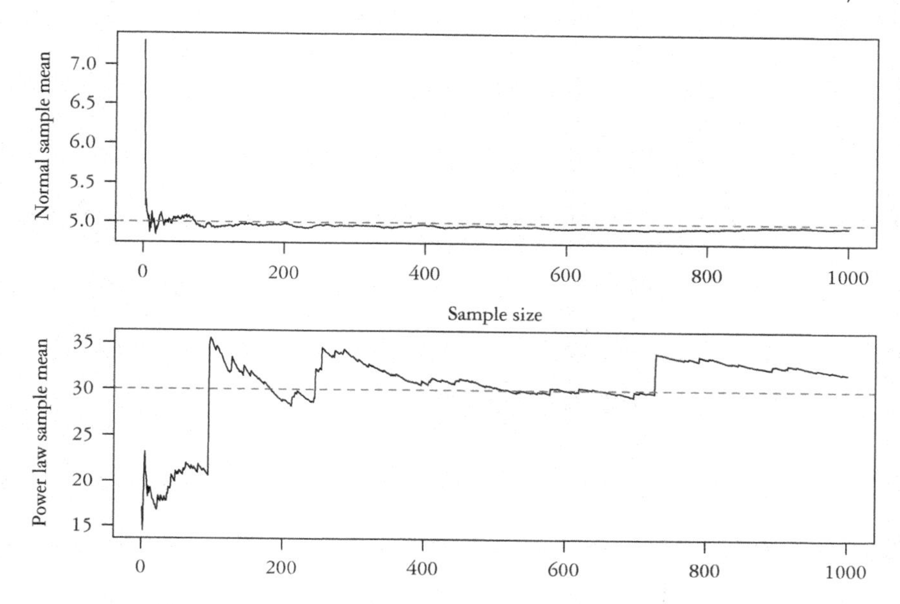

FIGURE 3.3 Running means of random samples drawn from normal (top) and power law (bottom) distributions as the number of observations increases from 1 to 1,000. Dashed gray lines represent the true average, or population mean.

We can see that in the case of the normal distribution, it does just that. The initial average of your sample observations may be off by a fair bit initially, but it very quickly converges to something close to the true underlying average value and more or less stays there. This is why we can get reasonable estimates of the characteristics of a population (on, say, a survey) just by looking at a sample.

The sample average of the data drawn from a power law distribution, by contrast, wanders all over the place and depends crucially on the number of large observations that occur in the sample. If there are more than we'd expect, the sample average will be too high. If there are fewer, it'll be too low. And because those large observations are really rare events, it takes an incredibly long time for them to average out and produce a sample mean that's reliably anywhere near the value that we're trying to estimate. That's why we can't just use standard statistical tests, like a difference-of-means test, to answer the question of whether war is becoming less deadly. Fortunately, as we'll see in chapter 5, there are some nonstandard statistical tests that can be of use.

Specifics aside, the point is this: The only way we can really know whether we're looking at a trend or at random noise is to rely on statistical tests to help us determine whether the number that we're trying to track, whatever it is, has changed more than we'd expect due to chance. Despite the fact that this problem is *at the heart of* the decline-of-war debate, such tests are routinely disregarded. In two recent articles in which there have been statistical tests of the battle-death data, the results indicate that the apparent trend is very plausibly an illusion—a normal pattern following the misleadingly large outliers of the late 1940s and early 1950s.[23]

Conclusion

This chapter and the previous one have laid out a handful of reasons that have prompted me to be skeptical of the decline-of-war thesis. Undermining the evidence in favor of that thesis is not, of course, the same thing as showing that there is no decline: The absence of evidence is not evidence of absence. It is much harder to answer the question of whether or not war is in decline, and if so, when, where, and why. I take up this challenge in the next three chapters.

PART III
What the Data Tell Us

4

Is International Conflict Going out of Style?

THE PRECEDING CHAPTERS HAVE highlighted some of my concerns about the metrics used in the decline-of-war literature. The next step is to take a careful look at the data in order to assess both the decline-of-war hypothesis and, to the extent possible, to evaluate the claim that countries have become less warlike over time.

This chapter starts off by asking what we mean, specifically, when we say that war is in decline. It describes three answers to that question—the rate of international conflict initiation, the deadliness of war, and the potency of the causes of war—that must each be measured in different ways. The rest of the chapter is devoted to measuring the first of the three, the rate of international conflict initiation over time, and exploring the question of whether and when it has changed in the past two centuries.

The answer, while not encouraging, is nevertheless intriguing. We don't see a generalized decrease in the rate of conflict initiation over time. If anything, we see *greater* rates of conflict initiation over time, at least until we come to the end of the Cold War. Those changes in the rate of conflict initiation are, by the standards of international conflict research, pretty substantial. At the same time, we see long periods during which the rate of conflict initiation does *not* change, which is also very interesting. The patterns of stability and change in the data give us important clues as to the forces that do drive conflict in the international system—a question that I pick up again in chapter 7.

Three Measures of Warfare

In chapter 2, I argued that some ways in which scholars have chosen to measure war are problematic. It didn't make much sense at that time to get into a deeper discussion of measurement and of which measures of war *are* appropriate, but before evaluating historical trends in warfare, it is obviously necessary to do so. Two questions need to be addressed: what should we measure, and how should we go about measuring it?

Measurement is tricky because the question of whether war is in decline is, itself, not very well specified. We could be asking whether conflicts happen less often than they once did, or whether they are less intense than they once were, or both. Indeed, as Professor Pinker (2011a, 210) argues,

> There is no single answer, because "warlike" can refer to two different things. It can refer to how likely people are to go to war, or it can refer to how many people are killed when they do. Imagine two rural counties with the same size population. One of them has a hundred teenage arsonists who delight in setting forest fires. But the forests are in isolated patches, so each fire dies out before doing much damage. The other county has just two arsonists, but its forests are connected, so that a small blaze is likely to spread, as they say, like wildfire. Which county has the worse forest fire problem? One could argue it either way. As far as the amount of reckless depravity is concerned, the first county is worse; as far as the risk of serious damage is concerned, the second is.

In other words, no single metric of warlikeness can be examined in order to evaluate the decline-of-war thesis. Because a decline of war could take the form of either a decline in the rate of interstate conflict *or* a decline in its intensity, we must examine both in order to assess the argument.

This position is utterly reasonable. The resort to violence is the essence of warlike behavior, especially since (as we have seen) a war, once begun, may grow far more deadly than any of the participants had anticipated. If the rate of conflict initiation in the international system is in decline, we should be able to detect that decline. Doing so is the main focus of this chapter.

It is also necessary to ask whether wars are fought with the same ferocity and savagery that they once were. If we have entered an era of more frequent but considerably less deadly wars, we might on balance conclude that war is in decline. Similarly, a decrease in the rate of war initiation would be less exciting if the wars that do occur are more deadly. For these reasons, I devote chapter 5 to an exploration of trends in the deadliness of war.

Finally, I examine a third set of trends: changes in the potency of the causes of war. If humanity is becoming less warlike, a territorial dispute today should be less likely to escalate to violence than a territorial dispute fifty or a hundred years ago would have been. More generally, if humanity is indeed turning toward pacifism, the conditions—material, ideological, territorial, what have you—that are associated with the onset of war should become more weakly associated with it over time. I explore this possibility in chapter 6.

Measuring International Conflict

There is good reason to think of the initiation of international conflict as the most important quantity to measure if we want to gauge trends in warfare. The decision by one country to use force against another, or of a sub-national group to declare open rebellion against the state, is a weighty one, in that it could result in the loss of a great many lives. Although leaders may underestimate just how deadly wars can become thanks to the dynamics of warfare, few are unaware of the potential. Moreover, given that the deadliness of war is very, very difficult to control without surrendering, a good case can be made that the initiation of conflict best reflects intentionality.

Measuring the initiation of international conflict is not without its difficulties, of course. Foremost among them is establishing a reasonable threshold beyond which violence can be said to have occurred. On the one hand, some low-level uses of force—drone strikes by the United States against the leaders of terrorist groups, for example—might seem extremely unlikely to escalate to warfare, at least in the conventional sense. At the same time, such judgments are often *post hoc:* They rely for their support on historical examples in which escalation to higher levels of violence did not occur, while ignoring the fact that it could well have done so. To continue the example, drone strikes have so far proven to represent a low risk of escalation, but were they to provoke a terrorist counterstrike on American soil the outcome might well be very different.

Because there is no obvious answer to the question of where to draw the line between peace (however strained) and armed conflict, students of international conflict have chosen to use different thresholds of violence when constructing their datasets. This outcome presents us with a great opportunity: Rather than trying to establish *a priori* which threshold is most legitimate, we can simply examine all of them.

The lowest level of conflict to be captured systematically by scholars is the *militarized interstate dispute,* data on which are collected by the Correlates of War project. The project, begun in 1963 by the political scientist J. David Singer at the University of Michigan, is an attempt to measure the objective, observable characteristics of international military conflicts. The source material for the original Correlates of War team members included such resources as *Keesing's Archive* and the *New York Times,* which were scoured by hand for conflicts dating back to the early 1800s. More recent iterations of the dataset have benefited from modern automated document classification techniques. Due to its thoroughness and depth of coverage, the Militarized Interstate Dispute dataset has become one of the most widely used datasets in the study of international conflict.[1]

In the words of three of the scholars who spearheaded the original data collection, militarized interstate disputes (or MIDs) are

> historical cases of conflict in which the threat, display or use of military force short of war by one member state is explicitly directed towards the government, official representatives, official forces, property, or territory of another state. Disputes are composed of incidents that range in intensity from threats to use force to actual combat short of war. (Jones, Bremer, and Singer, 1996, 163).

These three categories—the threat, display, and use of force—were chosen both because they collectively capture most forms of belligerence used by countries and because they represent a credible risk of escalation to war. Indeed, Professor Pinker (2011*a*, 282–283) himself argues that militarized interstate disputes "should be shaped by the same causes as the wars themselves, and thus can serve as a plentiful surrogate for wars."

That said, it could reasonably be argued that threats of force often carry virtually no risk of escalation. Prior to their development of thermonuclear warheads and ICBMs in 2017, for example, North Korea's impotent threats against the United States of annihilation and "final doom"[2] mostly provided fodder for late-night comedy shows. Displays of force, too, can be somewhat debatable: US Coast Guard vessels protecting fisheries from intrusion generally qualify as a display of force but seem very unlikely to go beyond a display.[3] Moreover, given that the goal is to measure the rate at which violent force is actually used by one country against another, an analysis that includes threats or displays of force would quite reasonably be open to the criticism that such events are irrelevant.

Actual uses of force, on the other hand, generally do involve the risk of escalation. These events encompass all uses of force, including not just wars but blockades, occupations, seizures of material or personnel, and other authorized uses of force by regular armed forces. To avoid skepticism over unauthorized cross-border incursions, random potshots across demilitarized zones, and the like, the scholars who created the dataset went to great pains to exclude trivial or irrelevant incidents: To qualify as a MID, an incident must be "an explicit, non-routine, and governmentally authorized action" (Jones, Bremer, and Singer, 1996, 169). Most important for our purposes, they represent an explicit decision by the officials of one government to pursue their goals by violent means.

Even so, some of these disputes are controversial in the conflict studies community because the use of force in question seems very unlikely to result in war. The most typical examples are fishery disputes involving seizures. For example, militarized interstate dispute #601 in the dataset records a use of force by Peru against the United States. This was a case in which an American fishing vessel, having strayed into Peru's territorial waters, was escorted to the port city of Talara by two Peruvian gunboats. Technically, the fishing boat had been seized by force, and the seizure lasted for more than twenty-four hours, so it meets the criteria for a militarized dispute. But it was released without any fines being assessed, and the American government never issued a protest of any kind. A similar set of cases involve the so-called Tanker War between Iraq and Iran during the 1980s. During that conflict, Iraq and Iran both fired on tankers in the Gulf, for different reasons—Iraq to damage Iran's ability to export oil, and Iran to intimidate other Gulf countries that were supporting Iraq. These actions are recorded as attacks on the country in which the ship was registered. Unfortunately, tanker registries are often decided based on operating costs like crew salary and the price of insurance, which can vary significantly from one country to the next. As a result, a tanker's flag very often does not reflect its country of ownership. Iraq and Iran are therefore recorded as having attacked the Bahamas, Libya, Cyprus, and a host of other countries during the conflict. To make matters even more confusing, the US Navy offered escort protection only to US-flagged vessels—which in late 1987 included 11 Kuwaiti-owned tankers that had hastily re-registered in the United States but excluded fifty-two American-owned tankers still flying foreign flags. In short, untangling who was attacking whom during this conflict is a real mess, and most of the countries "attacked" by Iran and Iraq were extremely unlikely to retaliate (Gibler, Miller, and Little, 2016, IA-5; Suro, 1987).

Professor Pinker made exactly this point about the 2013 conference paper that grew into this book. In that paper, I examined all international uses of force and argued that each had the potential to escalate to full-scale war. Responding to questions from journalist Anna Kordunsky (2013) for a story that appeared in *National Geographic*, Pinker responded that

> measuring all instances of conflict, without distinguishing between major wars and minor skirmishes, conflates "trivial shots across the bow" with wars that kill significant numbers of people. While Braumoeller refers to each instance as a "roll of the dice" with the potential to escalate into a larger war, Pinker argues that all interventions are not the same: there is far less escalation risk in firing cruise missiles into Sudan, as the U.S. did in 1998, than in invading Iraq, as the U.S. did in 2003.

I have to admit, I was a little annoyed at that criticism, given that Pinker had endorsed MIDs as useful proxies for war in his book. But it's a valid concern. While all forms of violence—threats, displays, or uses of force—might be relevant to the broader claim that the human proclivity to engage in violent behavior is in decline, some unreciprocated uses of force may represent situations in which neither side sees any meaningful risk of escalation to war.

The best way to avoid an endless series of arguments over US Coast Guard boats and cruise missiles and "shots across the bow," it seems to me, is to limit our investigation only to uses of force that were reciprocated by the target country. By so doing, we can categorically eliminate the concern that the target country was never really targeted (fishing disputes) or had no ability to respond in kind (Sudan) or wasn't willing to risk an escalated military response from the initiator. Simply put, reciprocated uses of force are evidence that both sides are willing and able to fight. Because "reciprocated international use of force" or "reciprocated level-4 militarized interstate dispute" gets pretty cumbersome (and "RL4MID" is just obnoxious), I'll just refer to these events as *international conflicts.*

Another measure of armed conflict that is worth examining is the result of a collaborative effort between the Conflict Data Project in the Department of Peace and Conflict Research at Uppsala University and the Conditions of War and Peace Program at the International Peace Research Institute, Oslo (PRIO). The UCDP/PRIO Armed Conflict Dataset (Gleditsch et al., 2002) is more conservative in its definition of a meaningful use of force. To qualify, a conflict must produce at least twenty-five battle deaths in a single year. The UCDP/PRIO dataset captures both international and civil conflicts. Unfortunately, at this writing the data only extend back to 1946.

The Correlates of War project also maintains data on a third measure of armed conflict: all-out wars, either interstate, intra-state (i.e., civil wars or intercommunal wars), or extra-state. Extra-state wars are wars between an established country and some entity that has not been formally recognized as a country, such as a colony. While the end of the colonial era understandably resulted in a decrease in the frequency of extra-state wars, the emergence of extra-state groups like al-Qaeda, the Taliban, and ISIS has ensured that the category remains relevant. All-out wars are understood to involve sustained combat, involving organized armed forces, resulting in a minimum of a thousand battle-related fatalities over a twelve-month period. To qualify as a participant in a war, a country must either sustain a minimum of one hundred battle-related fatalities or have a minimum of a thousand armed personnel engaged in active combat.[4] These data are also available back to 1816.

At what point should we say that the use of armed force has taken place? There are tradeoffs inherent in any answer. If we limit ourselves to only looking at all-out wars, we are almost certainly ignoring a lot of conflicts that could have escalated to all-out war but didn't. To the extent that superpower competition and nuclear weapons limited escalation during the Cold War, looking only at all-out war might fool us into believing that wars have become less common when in fact they have simply become more limited. Conversely, lower-level conflict might not represent the same societal commitment to the use of lethal force that all-out war does. Another tradeoff is inferential: It's a lot harder to spot patterns in data when you have less data, so only looking at all-out wars might hide results that would be visible if we examined lower-level conflict data, which afford us higher resolution.

My own take is that we're interested in officially sanctioned, reciprocated uses of force, however big or small, for a couple of reasons. First, the line between not using force and using force is much clearer and much more salient than the line between fighting a war that kills nine hundred people and a war that kills a thousand. If our goal is to gauge trends in the use of violent force in international relations, we can best accomplish it by drawing the line at the use of violent force. Second, as I've mentioned before, countries that use force against one another generally know that escalation is possible and the resulting bloodshed could be very substantial. Their decision to take that risk is the decision that interests me. Finally, the fact that there are many more international conflicts than thousand-death international wars means that trends in the former are a lot easier to spot.

That said, because other positions are not without their merits, and because the data are readily available, it makes sense to take a look at trends in conflict initiation at all three levels of violence—the reciprocated use of force, twenty-five deaths, and a thousand deaths. If those trends tell us very different stories, we'd have to make sense of them. But as we will see later in the chapter, the stories they tell are actually pretty consistent.

Measuring the Rate of Conflict Initiation

In order to measure the *rate* at which force is used by countries, we need to define not just the numerator but the denominator—that is, once we've figured out the number of cases in which force was used in a given time period, we need to figure out the number of cases in which force *could* have been used. This is exactly the procedure that Professor Pinker (2017) advocates: "it's only by (1) counting the violent incidents, (2) scaling them by the number of opportunities for violence to occur, and (3) seeing how this ratio changes over time that one can get an objective sense of trends in violence."

Measuring the number of opportunities for international conflict is a challenging task. Obviously, as my students figured out in the preface, we need to take the growth of the international system into account. As they also pointed out, correcting for that growth involves more than just counting the number of country pairs because not all pairs of states are equally capable of fighting one another. While Bolivia could easily attack Paraguay, for example, it would have a much harder time attacking Botswana. This issue becomes more acute over time because new states tend to be smaller, weaker, and farther apart than their predecessors, on average.

International relations scholars working in the quantitative tradition have long been aware of this issue. Our statistical models of war and peace don't work very well if they are tested on countries that cannot reach one another. To resolve this problem, Professors Bruce Russett and Zeev Maoz developed the concept of a "politically relevant dyad"—that is, a dyad, or pair of countries, in which the capabilities of at least one country are sufficient to overcome the distance between them. Put more simply, these are countries that could fight if they chose to do so. They laid out the following criteria for political relevance: Only neighboring pairs of countries, countries separated by less than 150 miles of water, and pairs of states that contain at least one major power are treated as potential combatants (Maoz and Russett, 1993).

While it represents a vast improvement over the assumption that all pairs of countries are equally relevant, this measure is admittedly coarse. Some non–Great Powers can fight across greater distances than others, and not all Great Powers have had truly global reach. Accordingly, Professor Austin Carson and I developed a more nuanced measure of political relevance by estimating rather than assuming the relationship between capabilities and distance, on the one hand, and political relevance on the other.[5] The resulting measure is slightly better than the Maoz-Russett measure at correctly sorting conflicts from non-conflicts.[6] Because the measure is continuous, it nicely captures the idea that political relevance is not an either-or proposition: Some pairs of countries are more politically relevant than others.

The political relevance of a pair of states reflects the degree to which they have the opportunity to fight. Powerful neighbors could fight very easily, while weaker states at a distance would be harder pressed to do so. The sum of the political relevance scores of all pairs of states in a given year, then, serves as a good measure of opportunity.[7] The measure of the rate of conflict initiation then becomes the number of international conflicts in a given year divided by the sum of the political relevance scores of all pairs of states in the system.

Trends in the Rate of Conflict Initiation

Of all indicators of warfare, trends in conflict initiation arguably hold pride of place. As I've noted previously, leaders who initiate armed conflict can never really be sure how far it will escalate. While many militarized disputes produce very few deaths, no one can be absolutely certain of that fact at the time, and everyone is aware of that fact. Authorizing the use of military force therefore represents a willingness, if things go poorly, to shed quite a lot of blood.

Our first glance at the data suggest that that willingness is indeed on the decline in recent years. Figure 4.1 examines trends in the rate of international conflict initiations that produced twenty-five or more battle-related fatalities from the end of the World War II through 2014. The horizontal red lines in this and the other figures in this chapter were estimated via a *change-point detection* algorithm, which does pretty much what the name suggests: It looks for changes in the distribution

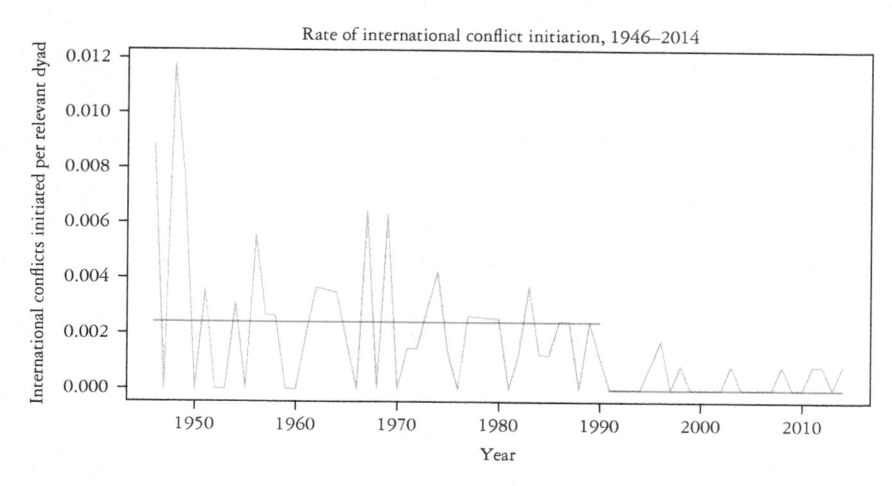

FIGURE 4.1 Change-point analysis of UCDP/PRIO international conflict data, 1946–2014. The light gray trend line charts the rate of international conflict initiation; the horizontal red lines indicate medians. Breaks between the horizontal lines indicate points at which the change in the overall trend is larger than one could reasonably expect by chance.

of some variable over time that are too large to attribute to chance. I use it here to pinpoint changes in the rate of international conflict initiation.[8] The horizontal red lines in each figure capture the median rate of conflict initiation over time. Breaks between horizontal lines indicate a change in the rate of conflict initiation that's too large to have occurred by chance. In this figure, the change-point algorithm finds a single change in the rate of conflict initiation, located precisely at the end of the Cold War. It finds no others. Based on these data, we can reasonably conclude that the average rate of international conflict initiation was constant throughout the Cold War, dropped promptly at the end of the Cold War, and has remained at its current level since.

That seems like unambiguously good news—and it is. Joshua Goldstein (2017), whose argument about the decline of war in *Winning the War on War* was limited entirely to the post–Cold War period, might very well be correct that the success of peacekeeping plays some part in this trend, though given the fact that most of those missions were sent to pacify civil rather than international wars I would expect to see

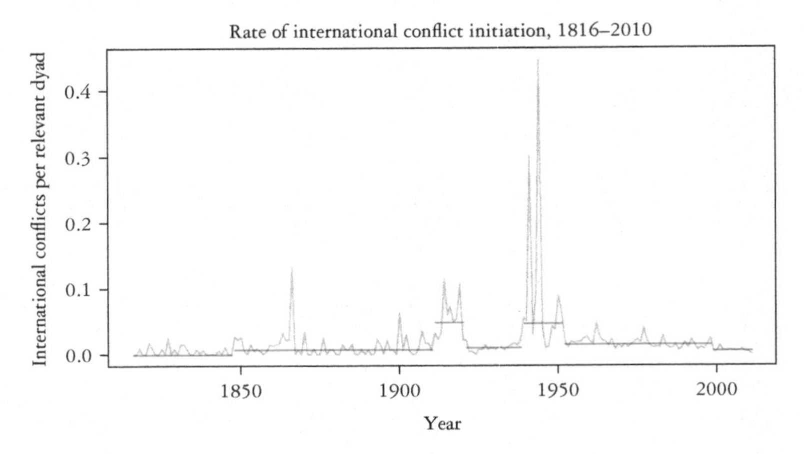

FIGURE 4.2 Change point analysis of Correlates of War use of force data, 1816–2010. The light gray trend line charts the rate at which force was used and reciprocated; the horizontal red lines indicate medians. Breaks between the horizontal lines indicate points at which the change in the overall trend is larger than one could reasonably expect by chance.

peacekeeping have a larger impact on trends in civil war initiation.[9] More likely, though, the change reflects the fact that one of the two superpowers that had been fighting proxy wars around the globe for the last forty-five years abruptly stopped fighting. Regardless of the reason, though, the numbers are pretty unambiguous: We live in a world that is considerably less plagued by international conflict than it was during the Cold War. That's a very good thing.

When we use reciprocated uses of force, for which we have data back to 1815, as the measure of international conflict, however, we're in for a pretty unpleasant surprise. Figure 4.2 charts the worldwide rate of international conflict onset, again weighted by the political relevance of the dyads that make up the system. We can see the two obvious spikes that represent World War I and World War II; we can also see a smaller spike around 1870, reflecting both the wars of German unification and a rash of wars in Latin America. Unfortunately, the spike in the rate of the use of force at the time of World War II is so shockingly high that it is difficult to compare trends outside of it. Accordingly, I have reproduced this figure as Figure 4.3, the only difference being that the Y-axis is shortened to make the trends more visible.

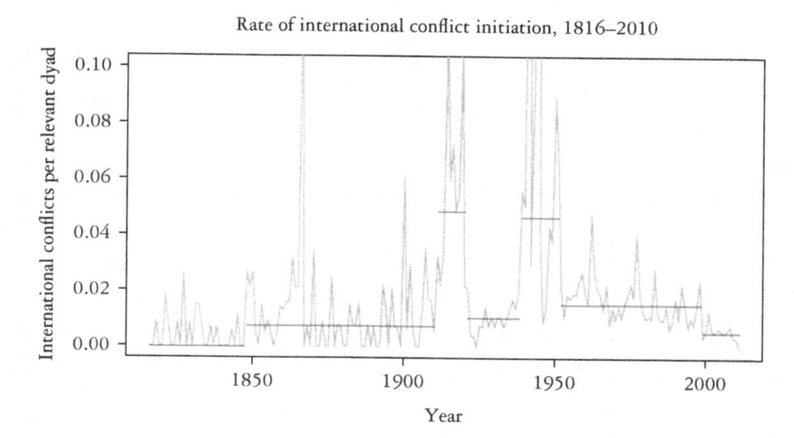

FIGURE 4.3 Change-point analysis of Correlates of War use of force data, 1816–2010, with Y-axis truncated to make trends more visible. The light gray trend line charts the rate at which force was used and reciprocated; the horizontal red lines indicate medians. Breaks between the horizontal lines indicate points at which the change in the overall trend is larger than one could reasonably expect by chance.

The story told by Figure 4.3 is pretty grim. Like Figure 4.1, it shows a significant drop around the end of the Cold War. The overall trend over the course of the past two centuries, however, has been an *increase* in the rate of conflict initiation between countries. In fact, if we leave out the two World Wars, we can see that the Cold War was the most conflictual peacetime period to have occurred since the Napoleonic Wars, and the end of the Cold War was the first instance of a decrease in the rate of conflict initiation in nearly two centuries. We can't know whether trends in conflict with twenty-five or more battle deaths follows the same pattern as international conflicts since 1816, simply because the UCDP/PRIO conflict data don't go back that far, but the fact that the trends that they chart during and after the Cold War correspond so closely at least suggests that that's a reasonable hypothesis.

Granted, most of these uses of force did not escalate into full-blown warfare. A determined critic might still argue that, if we are trying to say something about the human propensity to use violence against other humans, even reciprocated uses of force are too low a bar and that the real quantity of interest is the rate of initiation of all-out wars.

Frankly, I just don't buy this argument. For one thing, a use of force is *precisely* the quantity of interest: The decline-of-war arguments reviewed here argue that people are increasingly less willing to use force against one another, period. No one makes the case for an anti-violence norm that only kicks in when we reach 999 battle deaths. (One could argue that each new death is more objectionable than it would have been years or decades ago, but the way to test that claim is to look at war intensity or severity, not initiation.) For another thing, no one would have guessed that the self-immolation of a Tunisian street vendor named Tarek el-Tayeb Mohamed Bouazizi would lead to American, French, and Russian uses of force in Syria—and yet that is precisely what happened, thanks to the volatile political and social forces that translated Bouazizi's death into the Arab Spring. And the self-immolation of a Tunisian merchant is far less provocative than an officially authorized use of force by one country against another. Whether or not a use of force will escalate to all-out warfare is always easy to predict after the fact, but we just don't possess the sort of omniscience necessary to know which uses of force will escalate, and to what degree, at the time that force is first used. So for my money, the most interesting data from the point of view of the willingness of countries to use force against one another are the data in Figure 4.3, and those data offer little grounds for optimism.

Nevertheless, trends in all-out warfare are interesting in their own right, so it is worth examining them as well (Figure 4.4). Full-blown interstate wars are considerably more rare than uses of force, of course, so there is an obvious tradeoff in terms of the granularity of the data. Nevertheless, even this fuzzier picture tells a story that is difficult to square with the decline-of-war thesis. As far as all-out war is concerned, we can detect no decline after the Cold War, which was not appreciably more or less warlike than the interwar period, and we don't see anything like a steady decrease in the rate of war initiation over time.

One caveat is worth noting. On all of the change-point analysis plots in this book, I plot median lines through the data in each time period. I choose the median, or fiftieth percentile, because when distributions are skewed (as these are) the median is generally a better measure of the average tendency of the data than the mean, or simple average. That's

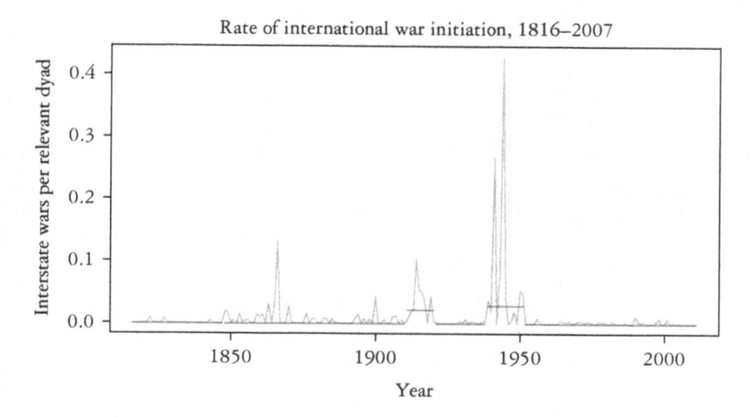

FIGURE 4.4 Change-point analysis of Correlates of War interstate war data, 1816–2007. The light gray trend line charts the rate at which interstate war occurs; the horizontal red lines indicate medians. Breaks between the horizontal lines indicate points at which the change in the overall trend is larger than one could reasonably expect by chance.

still true in Figure 4.4, but if you look carefully at the median lines around 1848 you can see a faint break. There *is* a significant change point at that time. We can't see it very well in this graph because wars are so rare: The median number of wars started in both periods is zero. If we zoom in to take a closer view and plot mean rather than median lines to compensate for the frequency of zeros in the data (Figure 4.5), we can see that, outside of World Wars I and II, the order from most to least warlike would be: second half of the nineteenth century; Cold War and post–Cold War periods; interwar period; and first half of the nineteenth century. That does mean that there was a decrease in the rate of war initiation after World War I. But it represents the only decrease in a two-hundred-year period that is otherwise characterized by a series of steady *increases*. When bellicosity is measured as the propensity to engage in all-out wars, not only was the Cold War the second most warlike period in the past two hundred years outside of the two World Wars, but there is no evidence that anything has changed since 1945.

Why doesn't the war initiation graph (Figure 4.5) look more like the conflict initiation graph (Figure 4.3)? They're actually not that dissimilar, if we note two things. First, the ratio of wars to conflicts

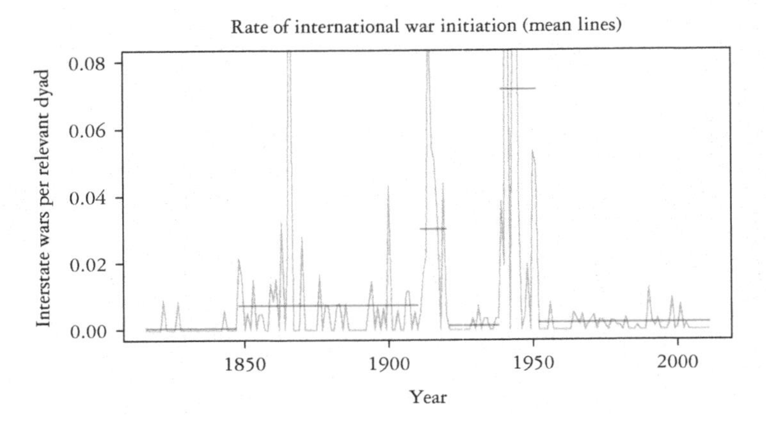

FIGURE 4.5 Change-point analysis of Correlates of War interstate war data, 1816–2007, with mean rather than median lines and Y-axis truncated to make trends more visible. The light gray trend line charts the rate at which interstate war occurs; the horizontal red lines indicate means. Breaks between the horizontal lines indicate points at which the change in the overall trend is larger than one could reasonably expect by chance.

was much higher during the second half of the nineteenth century than it was during any other period on record. That makes some sense to me because, to preview a point that I'll develop in much more depth in later chapters, this was the only period that lacked a general, functional international order that might deter conflicts or prevent them from escalating. The Concert of Europe had waned by that time, and the Treaty of Versailles had yet to be written. German Minister President Otto von Bismarck did implement such an order in the 1870s and 1880s, but only among the Great Powers. And the decrease in war and conflict initiation that it produced, while evident in the trend line, wasn't substantial enough to move the needle for the international system as a whole.

The second thing to note is that, when it comes to measuring bellicosity, wars are just a coarser measure than international conflicts. There's not a lot of variation in the rate of war initiation during and after the Cold War, mostly because wars in general are a pretty rare phenomenon. I am inclined to believe that the post–Cold War shift is real, despite the absence of a significant change in the rate of war initiation, because the rate of conflict initiation is a more sensitive measure. I also think it's a better measure in general, for the reasons I listed above.

Interpretation

No matter how we measure the rate or frequency of conflict or war initiation, these results are very difficult to square with the arguments, made by Professors Pinker and Mueller, that war is going out of style and has been for quite some time. Not even a "big dose of randomness" can square these data with Pinker's claim that we've seen an overall decline in warfare for centuries. Random fluctuations in the rate of conflict initiation over time show that there is, indeed, quite a bit of variation from one year to the next. But once we take that randomness into account we see a steady increase, rather than a decrease, in the rate at which international conflict occurs across almost two centuries.

The results are also difficult to square with Professor Goldstein's (2011, 238) argument that peacekeeping is responsible for the lion's share of the reduction in conflict in recent years. To be clear, Goldstein (2011, 43) does credit the end of the Cold War for much of the effect:

> [t]he end of the Cold War certainly helps explain the big reduction in war violence after 1989. During the Cold War era, each superpower provided support to governments or rebels in proxy wars in Asia, Africa, and Latin America.... Regardless of whether you blame the Soviet or American side for these proxy wars, the simple fact is that the superpowers' involvement in dozens of them around the world greatly increased the level of war violence during the Cold War years.

Still, Goldstein (2011, 43) argues, the end of the Cold War does not explain "the decreasing levels of overall violence in the 1970s and '80s [or] the continuing decline in violence over the past twenty years, after the Cold War ended." The above analysis suggests that no such explanation is necessary, at least in the case of interstate conflict: No such decreases are evident, no matter which measure is used.

That said, there is variation in the rate of conflict and war initiation over time, and it's pretty substantial. Leaving aside the two jumps during the World Wars, the median rate of conflict initiation *quadruples* in the period between 1815 and the end of the Cold War, after which it abruptly drops by more than half. By the standards of international conflict scholarship, those are huge, tectonic changes. Those of us who do this sort of thing for a living are usually content to find a variable that produces a change of a few percentage points in the probability of conflict initiation, as long as the change is statistically significant. From that perspective, finding changes of this magnitude in the rate of conflict initiation brings to mind one of those cartoons in which a fisherman, impressed with the size of the fish he's just caught, looks down and realizes that his boat is resting on the back of a whale.

The technique I've used to come to these conclusions is designed to let the data speak. The breaks between the horizontal lines represent the most likely breaks between periods that are characterized by different levels of conflict, as determined by a very flexible and powerful statistical algorithm. It doesn't impose any parametric assumptions on the data at all, which mostly means that I'm not assuming that they conform to any particular distribution when coming up with these estimates. That's important because, given how statistical techniques work, those assumptions can do some heavy lifting that ideally should be done by the data themselves. Here, the data are free to tell the story, and the story tells us that variation in the rate of conflict initiation over time isn't just random—in fact, it's huge.

It's also important to note that, over fairly long stretches of time, the rate of conflict initiation *doesn't* change. The way I've set the parameters, the technique is capable of picking up changes across periods that are as short as a decade in duration,[10] and looking at the two World War periods we can see that it does so. But there are also surprisingly long periods during which we don't see a change. Only one change occurs in the whole of the nineteenth century, for example, and by most measures there's no change at all during the Cold War.

The absence of change is as interesting as the changes that occur. Together, what they tell us is that if we want to understand the cause of these changes in the rate of international conflict initiation, we need to be looking for something that can change quite dramatically but also *doesn't* change over fairly long stretches of time. That's not a pattern that, to me at least, immediately suggests that we should look to (typically more gradual) evolution in the beliefs or ideas of individual human beings as the primary driver of change.

I have, I think, a very good idea of where we should look: to the level of the international system as a whole, and in particular, to international order. The patterns of stability and change are clues that, to someone who's read even a little about international history, jump right off the page. It's hard to miss the absence of change throughout the Cold War, for example, or the abrupt dropoff at the end. European states made up a much larger percentage of all states in the international system in the early 1800s than they do today, and the extraordinary period of calm at the beginning of the period corresponds to the period during which the Concert of Europe guided, to some degree, the decisions of European leaders.

Fleshing out this perspective on the data, and doing additional tests to explore it further, are the goals of the third section of this book, so I'll leave off the discussion here and pick it up in chapter 7. For now, I'll be content to note that there are changes in the rate of conflict of war initiation across historical periods and those changes are *very* large, but they don't at all look like what we'd expect to see if the gradual improvement of humanity were responsible for changes in the rate of international conflict initiation.

But But But...

To my mind, the results just described provide the best available answers to the question of whether or not war is in decline. Of course, reasonable people might have different ideas about which tests are best or most appropriate, or other questions about the data analysis. I'll deal with the most likely objections here and relegate my discussions of the rest to the Appendix.

Are the Data Reliable?

One issue that comes up when I discuss these findings is the possibility that a lot of older militarized interstate disputes were not recorded and did not show up in the data. To the extent that this is the case, the nineteenth century might appear more peaceful than it really was because we don't have a complete record of nineteenth-century conflicts.

There are a few reasons to be skeptical of this claim. First, the data only include level 4 or 5 militarized disputes—uses of force by one government against another. Lower-level disputes, such as threats and displays of force, are those most likely to be missed in previous centuries. Moreover, even a quick glance at the sources used by the Correlates of War project[11] for coding militarized interstate disputes reveals an impressively comprehensive array of source materials: Far from just relying on the *New York Times* and Facts on File, the Correlates of War project has delved into an impressive array of histories, biographies, notes, memoranda, military and political encyclopedias, and foreign-language sources. It's possible for an armed clash to go unrecorded in *any* of these sources, of course, but the scholars at the Correlates of War project have certainly done due diligence in minimizing that possibility. Finally, as chapter 8 will make clear, the trend in the data is very consistent with the mainstream historical record, which portrays the nineteenth century as a remarkably peaceful one.

Let's say, though, that these arguments aren't convincing and that we remain really worried about the possibility that the results are just an artifact of older militarized disputes not being recorded. The question then becomes, how many militarized disputes would have to have been missed for us to believe that the world is getting more peaceful? A good, rough-and-ready way to answer this question is to add hypothetical conflicts to the most conflictual part of the nineteenth century—that is, the second half—until the change-point algorithm tells us that its rate of conflict is significantly *greater* than that of the Cold War. When I did so, I found that the median rate of conflict would have to be about 2.5 times as high as its historical value before we could reasonably believe that peace broke out during the Cold War. In other words, the Correlates of War project would have had to have missed about *three uses of force out of every five* for this argument to hold water. That's a *lot* of missing data—so much so that I find it to be an awfully implausible claim. And since the first half of the nineteenth century is the more peaceful of the two, this is the *best*-case scenario for anyone who wants to claim that missing data are masking a real decline in the rate of the use of force.

What About Unreciprocated Uses of Force?

I decided in favor of using reciprocated uses of force as a measure of international conflict initiation above because scholars have argued that some militarized interstate disputes—fishing disputes, for example, or attacks on oil tankers flying the flag of another country—don't really represent any serious risk of military escalation. In doing so, I was being exceptionally cautious: Many existing studies of international conflict don't include the requirement that the use of force be reciprocated. It's entirely fair to wonder whether the answers change if we include unreciprocated uses of force.

As Figure 4.6 demonstrates, the overall picture doesn't change much at all. Trends in the rate of the use of force, reciprocated or not, do differ slightly from trends in the reciprocated use of force: The nineteenth century doesn't look quite as peaceful relative to other periods as it does when we only examine reciprocated uses of force. More striking, the early Cold War period—prior to the Cuban Missile Crisis—looks significantly more conflict-prone than it had previously. The missile crisis was a watershed moment in the Cold War, of course: The

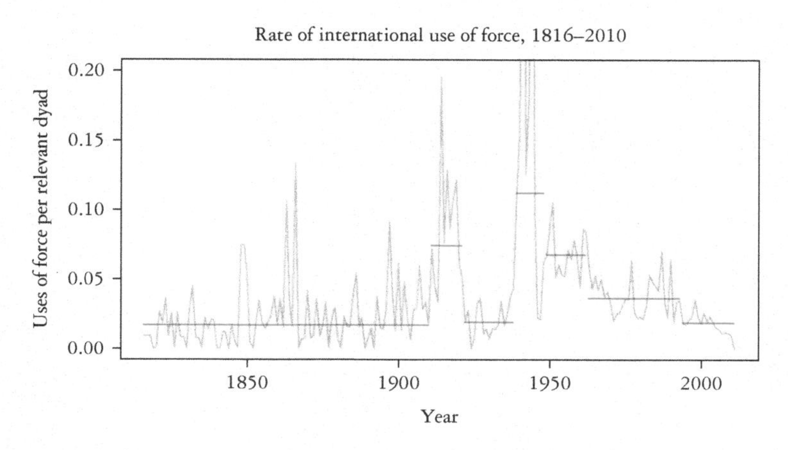

FIGURE 4.6 Change-point analysis of Correlates of War use of force data, 1816–2010, for both reciprocated and unreciprocated uses of force.

realization that the two superpowers had very nearly fought a nuclear war prompted a host of confidence-building measures and treaties as well as the installation of the Moscow–Washington hotline. So it's not unreasonable to think that the post-1962 world was a significantly more cautious place.

All told, the data on reciprocated and unreciprocated uses of force still don't support the claim that there has been a steady decline in warfare for hundreds of years. They do support, perhaps even to a greater degree, the conclusion that I offered earlier: the Cold War was an unusually violent period, and the post–Cold War drop in the use of force looks more like a return to normal levels of international violence than it does the culmination of a centuries-long trend toward a more peaceful world.

What About Other Kinds of International Wars?

There's another plausible reason to question the findings in this chapter: the interstate system in the nineteenth and early twentieth centuries didn't cover the whole globe. Interstate wars, therefore, do not include *extra-state* wars (wars between states and nonstate actors, like colonies) and *nonstate* wars (wars between nonstate actors), which happened a fair bit during that time. To the extent that the form that warfare takes has changed over the last two hundred years, those changes may obscure our ability to see larger trends.

Before proceeding further, I should offer a mild rejoinder. To my mind, the rate of international conflict is interesting in its own right. The causes of international conflict are overwhelmingly seen in the political science literature as being different than those of colonial conflict or nonstate conflict, much as homicides are seen as having different causes than death by lung cancer or death by sharkbite. My own argument about why the rate of international conflict does change, which I'll describe in later chapters, pertains exclusively to international conflict. And international conflicts kill a lot of people.

That said, there's a larger and much more practical objection that makes it difficult if not impossible to expand our definition of war to include these other kinds of war and then say something meaningful about changes in the rate of conflict initiation. That problem is not an absence of data on war: The Correlates of War project has also produced datasets on both extra-state wars and nonstate wars. The problem is that, in order to calculate the *rate* of war initiation in all such cases, we would need to measure the number of *opportunities* for war that each state had. We can do that in the case of interstate wars by adjusting for the political relevance of other states in the system, thereby avoiding counting Bolivia and Botswana as potential combatants. There is no equivalent technique to use for extra-state wars, much less wars among nonstate actors. There is simply no obvious, uncontroversial way to come up with an aggregate rate of conflict initiation given the multiplicity of peoples outside of the international system and the vagaries of the systems of governance under which they lived.

We can, however, see what the data look like if we look at the frequency rather than the rate of war initiation—that is, if we assume that the number of conflict opportunities hasn't really changed much over time. This amounts to assuming that, for every Kenya or India or Australia that Great Britain could potentially fight in the twenty-first century, there was one equivalent nonstate entity that Great Britain could have fought in the nineteenth century. And for every Botswana that Namibia could fight in the twenty-first century, there was one equivalent nonstate entity that the Bantu peoples and their fellow pre-Namibians could have fought in the nineteenth century.

You need only think about the number of wars that the British *could* have fought in the nineteenth century to realize that this assumption almost certainly understates the number of opportunities for conflict. It's true that the Great Powers didn't quite have the global reach then that they do now, but by that time they already had enough of a reach to fight wars in very far-flung parts of the planet. At the same time, in reality there were *vastly* more regions to encroach upon and peoples to fight then than there are now: Prior to independence, India alone was made up of no fewer than 565 princely states,[12] any one of which could hypothetically have fought the British and produced a war in the Correlates of War extra-state war dataset.

Because the number of opportunities for conflict in the nineteenth century is almost certainly understated, using the raw frequency of warfare will almost certainly make the nineteenth century look artificially warlike. If we somehow managed to include all of the additional conflict opportunities in the nineteenth century, the result would be a significant decrease in the rate of conflict onset during that period. Therefore, to the extent that this measure is biased in earlier periods, its bias should favor the decline-of-war thesis.

When we put all the data together, we can see that pooling the wars[13] fought against and among nonstate entities with the wars fought by states and measuring their frequency over time (Figure 4.7) gives us little reason to believe the decline-of-war thesis—not even after the end of the Cold War. In fact, this exercise highlights a point that's been missed in the decline of war literature: *extra-state wars are not all colonial wars.* Professor Pinker (2011*a*, 302) refers to "a kind of war that has vanished off the face of the earth: the extrastate or colonial war," but they're not the same thing. Insurgents and terrorist groups are nonstate actors too, and they've been involved in a lot of conflicts lately. In fact, an upsurge of extra-state war initiations in recent years—against the Afghan resistance in 2001 and the Iraqi resistance in 2003–2004, especially—may herald a renaissance of a form of warfare that had nearly died out after the dissolution of the European colonial empires. The ongoing

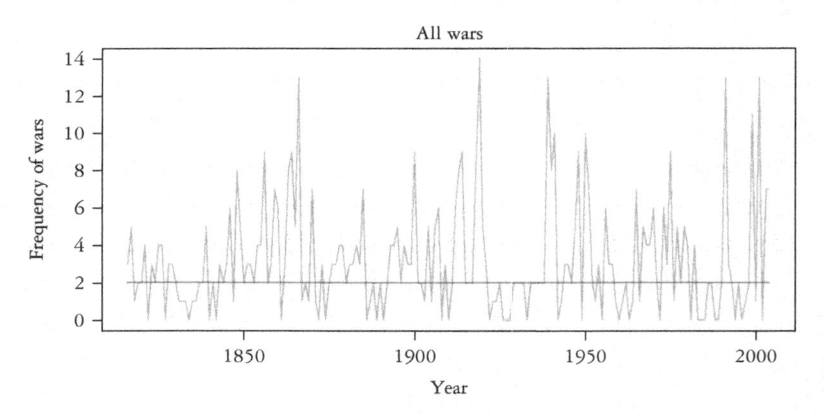

FIGURE 4.7 The frequency of war, 1815–2007. The time series represents the number of international, extra-state, or nonstate wars started in each year, according to the Correlates of War project data. The horizontal red line reflects the median. The fact that there are no breaks in the red line indicates that the change-point algorithm found no upward or downward trend in the data throughout the entire period.

war against ISIS is too recent to be captured in these data, but ISIS is considered to be a nonstate actor by everyone except ISIS, so it would not surprise me if this revival of extra-state warfare were to continue.

What About...?

Other questions might well have occurred to you as you read this chapter, especially if you're a specialist in this area. What about civil wars? What about other measures of war? What about other measures of political relevance? Other people have suggested a Poisson test for use on these data—what happens if you do that?

If you thought of questions like these, congratulations—you are a bona fide, hard-core nerd! And because you're a nerd, you probably don't mind looking up the answers to questions like these in an Appendix. Have at it—the relevant section starts on page 225. The other nice people here are probably getting a bit impatient to move on to the question of the deadliness of war, and I should accommodate them.

Conclusion

I have to admit that when I started this project a big part of me hoped that I would be able to show conclusively that conflict is in decline. I devoted my life to the study of conflict in the hopes that understanding conflict would improve the prospects for peace. I'm well aware that objectivity is essential to scientific inquiry, of course, but I also think we should be clear about our biases. My bias in this case is that I want to believe that the world has become more peaceful over time and will continue to do so.

Unfortunately, nothing in the data gives me much reason to sustain that hope. The rate at which countries use force against one another has increased more than it has decreased over the last two hundred years. The decrease following the end of the Cold War, while real, is the exception rather than the rule. The most exciting and intriguing finding is that we do see very substantial changes in the rate of international conflict initiation over time—a fact that might help us get a handle on the question of how we might change it in the future.

5

Is International Conflict Getting Less Deadly?

THE LAST CHAPTER FOUND no evidence of a downward trend in the rate of international conflict onset over time. Conflict onset, however, is only one measure of war. Another, equally interesting, is the deadliness of war, which Professor Pinker argues has been on the decline since 1945. This chapter examines that claim.

What it finds is really not uplifting at all. Analyzing the two most commonly used measures of the deadliness of war, I find no significant change in war's lethality. If anything, the data indicate a very modest *increase* in lethality, but that increase could very easily be due to chance. The data do indicate a slight but significant change in a third measure—battle deaths as a percentage of world population—in small to medium-sized wars, but that change could easily be an artifact of the tendency of countries' militaries to grow more slowly than their populations. Most importantly, that difference *doesn't* appear in the deadly thick tails of the distribution, where the most shockingly lethal wars are found.

Worse still, the data are consistent with a process by which only random chance prevents small wars from escalating into very, very big ones. This escalatory process is shocking in its magnitude: Very few "black swan" phenomena escalate in size as readily as wars do. That doesn't mean that we can't choose whether to escalate or end a war, of course. But it does suggest that the choices we make in war, and the outcomes of those choices, won't be predictable *a priori.* If that's true, the only difference between a small war and a very, very large one is the chance events that influence escalation.

These conclusions are, by a wide margin, the worst news that this book has to convey. The past two centuries have seen some of the deadliest wars in human history, including the deadliest of them all, World War II. The data give us no reason to believe that the next two centuries will be any better, and they could easily be much, much worse. If chance events are the main drivers of escalation, anyone who starts a war today is running a small but nontrivial risk that the war will snowball to nightmarish proportions. And decision makers continue to roll the dice.

Measuring the Deadliness of War

While reasonable scholars differ, as we saw in the last chapter, on the threshold above which a significant use of military force can be said to have occurred, the same is not true of measures of deadliness. Almost without exception, the deadliness of war is measured in battle deaths. Battle deaths generally include soldiers killed by the armed forces of another country but do not include civilians, let alone indirect deaths due to starvation, disease, or displacement. While in principle it might be better to measure the total number of deaths,[1] doing so is exceedingly difficult, even today: Estimates of mortality are generally based on a comparison of aggregate population figures before and after a conflict, and as Professor Goldstein (2011, 260–263) points out, those figures can be dramatically misleading. While scholars have recently sought to obtain better and more comprehensive estimates, they have only been able to do so in a very limited number of cases.[2] Battle deaths are more straightforward to measure and serve as a useful proxy for the overall deadliness of a war.[3] The Correlates of War project measures battle deaths for wars going back to 1816, while the UCDP/PRIO data measure battle deaths for conflicts dating back to 1946.

While there is widespread agreement on the numerator when it comes to the deadliness of war, there is less consensus on the denominator. Not long after the inception of the Correlates of War project, Professors Melvin Small and J. David Singer (1982, ch. 3) derived three separate indicators of the seriousness of war:

- *Magnitude.* The total number of country-years spent waging a given war.
- *Severity.* The total number of battle deaths resulting from a war.
- *Intensity.* The total number of battle deaths resulting from a war, divided by the pooled prewar population of the combatants.

In subsequent years these measures have become well established in the quantitative study of international relations.[4]

Magnitude, clearly, is a measure of overall duration rather than deadliness, so it is not well suited for our purposes. Severity, on the other hand, is a direct measure of deadliness, meant to capture the human cost of a war. If we are interested in the question of which conflicts caused the largest number of battle deaths in absolute terms, looking at severity makes sense. The third measure, intensity, is meant to capture the *relative* cost of war to the societies involved. It is designed to account for the ability of larger societies to sustain larger losses as well as to control for the growth of populations over time.

As I mentioned in chapter 2, yet another school of thought suggests that the lethality of war should be measured, not as battle deaths divided by the prewar population of the combatants, but rather as battle deaths divided by the population of the entire globe—a measure that I'll call the *prevalence* of war. For the most part, this isn't a measure that's used in the statistical study of war, but it's been used fairly often in the decline-of-war literature.

It's interesting, and a little surprising, that there's no clear consensus on which measure to use. I think the reason for that lack of consensus is that some people think about war as a form of human behavior, while others think about it as a cause of human mortality. The people who think about it as a form of human behavior tend to focus on the phenomenon itself and its outcomes—such and such a war killed so many people. Those who think of it as a cause of mortality reflexively frame those deaths in such a way that they can be compared to deaths from other causes. Neither answer is right or wrong, but each suggests an imperfect measure for our purposes. Severity is insensitive to the fact that the societies that fight grow larger over time, which is probably not irrelevant to the absolute number of people who get killed in a war.

Prevalence counts everyone as a potential casualty of war in every given year, when in fact most of us aren't.

For these reasons, intensity strikes me as a useful middle ground. It's not a perfect measure, but it avoids the obvious criticisms of the others, and its results can be interpreted either as population-adjusted severity or as the magnitude of war as a cause of mortality for the people whose countries are actually involved in war. Because reasonable people differ, and for the sake of thoroughness, I'll also examine trends in severity and prevalence and see what we can learn from them.

I'll also look into an explanation for why wars as a percentage of population size get smaller over time that might be confounding these results. States with larger populations have smaller militaries as a percentage of their population, and those smaller militaries produce fewer war fatalities. Professor Rahul C. Oka (2017) and his colleagues and Professors Dean Falk and Charles Hildebolt (2017), in papers published just two days apart, find robust historical support for this argument. Larger societies aren't less violent: Smaller societies are more vulnerable and have to hedge more against the threat of war. Their larger militaries, in relative terms, produce more deaths, again in relative terms, than those of larger societies. As population grows, military size relative to population decreases, and the deadliness of war decreases—not because people are less warlike, but because militaries don't need to be as big to provide defense. That fact raises the possibility of a spurious negative correlation between time and the deadliness of war when per capita measures are used. In fact, that's exactly what I find in the case of war prevalence.

I use more statistical tests in this chapter than in any other, just because, to put it bluntly, the data are a real pain. The severity and deadliness of war follow a power-law distribution, which I discussed briefly at the end of chapter 3 and will discuss in more detail below. Standard statistical tests break when they're applied to data that follow a power-law distribution, so we can't solely rely on the change-point algorithm used in the previous chapter.[5] The best tests to use are those that make the most of the information in the thick tails of the distribution, where key insights about war's escalatory potential can be found. I explain them all as I go, but if you want a quick primer on how they work before you dive in, I'd urge you to check out the online Appendix. (The examples I explore there are kind of fun in their own right.)

What We Already Know: Wars Got Deadlier 250 Years Ago

As it happens, we already know something about changes in the deadliness of warfare over time. In a 2011 article, political scientists Lars-Erik Cederman and T. Camber Warren and physicist Didier Sornette (2011) found that the severity of warfare had actually *increased* at the time of the Napoleonic Wars. It's worth discussing the way they came to this conclusion, as I'll be using a similar methodology below.

First, the authors needed to figure out whether there had been a change in the distribution of war fatalities over time and, if so, about when that change happened. They made a list of all major-power wars since the War of the League of Venice in 1495, in order. Then they went down the list, and at each gap between wars they ran a statistical test to assess whether the deadliness of the wars that came before differed systematically from the deadliness of the wars that came after. To separate systematic differences from random noise, they used a very general statistical test, the Kolmogorov-Smirnov test, which is designed to flag systematic differences in distributions of data. They located one such point, which occurred around the time of the Napoleonic Wars.

To get a better handle on what had happened at that time, they split the data into pre-1789 and post-1789 wars. They then plotted the severity of each war (on the x-axis of Figure 5.1) against the probability of observing another war that's at least as large as that one (on the y-axis). Because the distribution of war fatalities is really, really skewed, with lots of minor wars and a few really big ones, plotting these variables on logarithmic (1, 10, 100, ...) rather than linear (1, 2, 3, ...) scales makes it easier to see the relationship between them. It also highlights the straight-line relationships in log-log space that are characteristic of power-law distributions.

Power-law distributions are the key characteristic of the highly improbable, unimaginably large phenomena that Nassim Taleb described in his bestselling book *The Black Swan.* If an outcome follows a power-law distribution, the overwhelming majority of the observations of that outcome are small, but a few are really, really huge.

Earthquakes, city sizes, and war fatalities are all examples of phenomena that plausibly follow a power-law distribution.

When someone shows you a logarithmic plot of data that follows a power-law distribution, as I'm about to do, the key thing to look for is the *slope of the line* that's drawn through the data. Shallower lines represent situations in which escalation happens easily and large observations are relatively more common, while steeper lines represent less escalatory situations in which large observations are comparatively rare. They'll also mention the *slope coefficient*, α, which comes from the mathematical formula for a power law: $\Pr(X>x) \sim x^{-\alpha}$. The α coefficient connects an outcome like war size to the relative probability of that outcome: A war that's twice as large as some other war, for example, is 2^{α} times as rare. So if $\alpha = 3$, wars that kill two hundred thousand people will be eight times as rare as a war that kills a hundred thousand people, but if $\alpha = 2$ they would only be four times as rare (Cederman, Warren, and Sornette, 2011, 621). Specifics aside, the key thing to remember is that lower values of α correspond to shallower slopes and therefore to deadlier, more escalatory wars.

As Figure 5.1 demonstrates, the slope of the line that describes the relative frequencies of large wars is considerably shallower after 1789

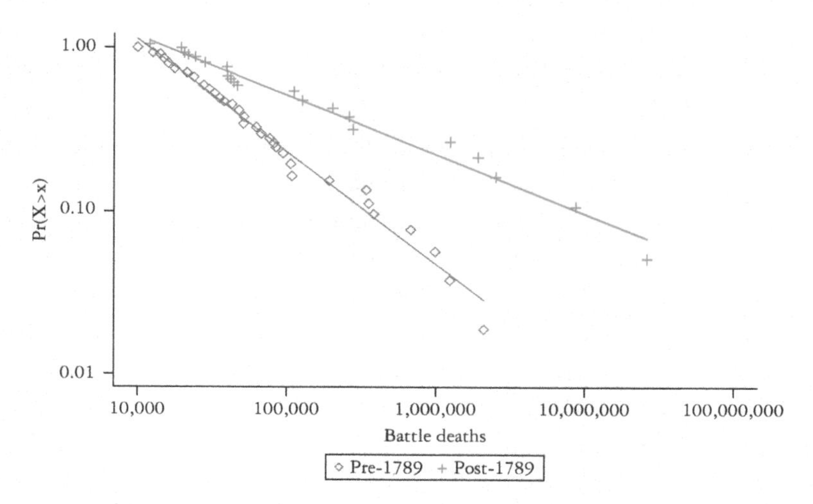

FIGURE 5.1 The increase in the severity of war following the Napoleonic Wars, as indicated by the shallower slope of the post-1789 line (Cederman, Warren, and Sornette, 2011).

than it had been previously, indicating an increase in the deadliness of war. As the authors put it, "[t]he substantial decrease in α for the latter period is strong evidence for a deep shift in the generating process of interstate war sizes, which inclined the global interstate system to produce wars of significantly greater severity after 1789." The authors also run a test to ascertain whether the difference between the two slope coefficients could plausibly be written off as chance variation in the data and find that it cannot.[6] The evidence is quite clear, then, that the deadliest wars were significantly deadlier after the Napoleonic Wars than they had been before.[7]

While this result doesn't address the claims of the decline-of-war theorists—popular perceptions to the contrary notwithstanding, even Professor Pinker concedes that the pre-1945 trend was toward *deadlier* wars—it is nevertheless interesting in its own right and as a template for the tests that follow.

Wars Haven't Gotten Less Intense Since Then

Our first look at the data on war intensity (Figure 5.2) is not encouraging. When the lethality of war is normalized by the population of the combatants and the wars are ordered sequentially by midpoint, it is sobering to realize that we can no longer reliably pick out World Wars I and II. Five wars stand out from the rest in terms of lethality, but without a concrete date along the x-axis to guide us it is difficult to know which is which. (Can you guess? I'll reveal the answer in a moment.) The similarity of battle deaths to population size across these five wars means that, regardless of which is which, there have been three other wars in the past two centuries that rivaled the World Wars in terms of intensity. That fact in and of itself is jarring.

The figure becomes even more sobering when we consult the mortality data and realize that the two World Wars are ranked fourth and fifth in the top five. (They are the second and fourth spike from the left, respectively, of the tallest five.) That means that three local wars—two involving the unfortunate country of Paraguay—actually exceeded the mindboggling savagery of the two World Wars. The first, in chronological order, was the Paraguayan War, a disastrous war that pitted Paraguay against Argentina, Brazil, and Uruguay in the 1860s

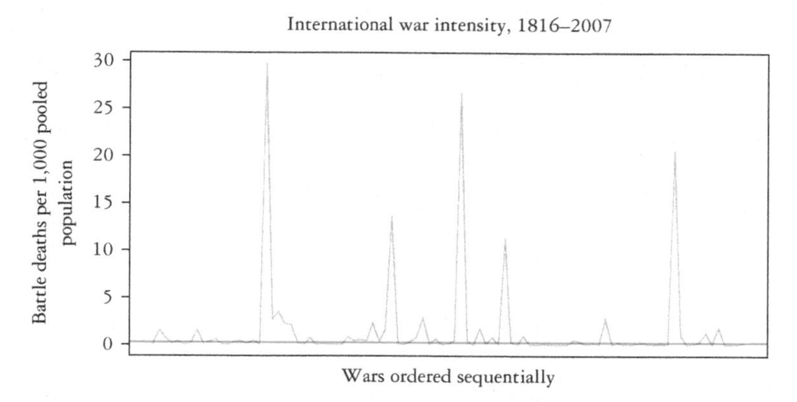

FIGURE 5.2 War intensity (battle deaths divided by pooled combatant population), with results from the change-point algorithm indicating no change over time.

and resulted in the deaths of something like 70% of the adult male population of Paraguay. The second was the Chaco War, a vicious conflict between Paraguay and Bolivia in the 1930s over suspected oil reserves. Paraguay won, again at great cost, but they failed to find any oil for the next seventy years.[8] The third was not Vietnam or Korea but rather the Iran-Iraq War in the 1980s, a bloody and protracted conflict that resembled World War I in the tactics and weapons used (trenches, barbed wire, poison gas) but actually exceeded it in intensity.

These three conflicts underscore the rationale for using intensity as a measure of war's lethality. They are not common subjects of history books, nor are they likely to be the backdrop of a Hollywood blockbuster. Many readers will probably not have heard of at least two of them (I hadn't, before I started studying war for a living). The body counts are nowhere near those of World War I or II in absolute terms. But no sane person would want to be caught in the middle of one of them. The residents of Iran and Iraq in the 1980s, or of Paraguay and Bolivia in the 1930s, did not think of their losses as being small relative to the world's population—they thought of them as being horrifyingly large relative to their own. And as I pointed out in chapter 3, if we are attempting to understand the decision to use violence, we should measure violence as the people themselves perceive it.

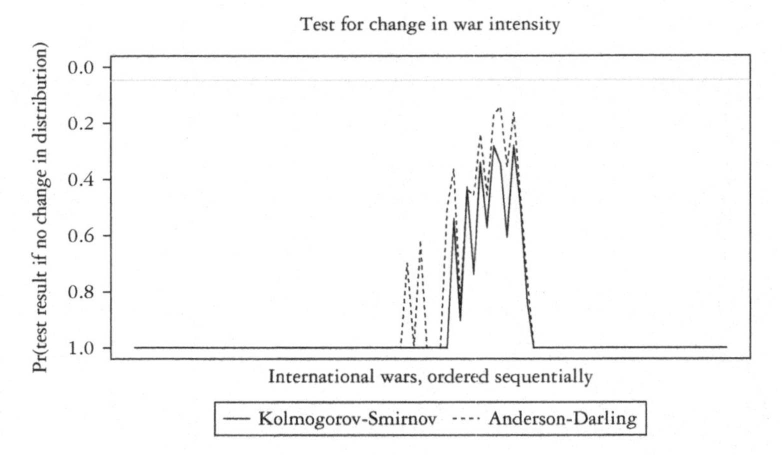

FIGURE 5.3 Results from Kolmogorov-Smirnov and Anderson-Darling tests for changes in the distribution of war intensity. Values on the y-axis closer to zero (note the reversed axis) indicate that the wars on either side of the cut point are less likely to have the same distribution. Because neither the solid nor the dotted line crosses the gray horizontal line at $y = 0.05$, we cannot conclude that a change in the distribution of war intensity has taken place.

As the red line in Figure 5.2 suggests, the change-point algorithm finds no particular upward or downward trend in the data over time. Some stretches without especially intense wars are longer than others, but of course that's what random variation looks like. And wars are rare enough that we cannot with any confidence conclude that wars as intense and bloody as these are a thing of the past.

What happens if, like Cederman, Warren, and Sornette, we use the Kolmogorov-Smirnov test to look for points in history at which the distribution of war intensity has changed? This simple little test, one of the most useful tools in any statistician's toolbox, is designed to measure the probability that two samples were drawn from the same underlying population. In Figure 5.3, I use the Kolmogorov-Smirnov test to look for points in time at which we might reasonably conclude that there *has* been a change in war intensity. For the sake of thoroughness, I also include the results of the Anderson-Darling test (dashed line), a test that's designed to do the same thing but is more robust to differences in the tails of distributions.[9]

In Figure 5.3—as in all plots of the results of these two tests—the wars are again ordered chronologically along the x-axis. The height of the lines reflects the probability, based on these two statistical tests, that wars in the earlier period were neither more nor less deadly than they were in the other. Some variation in the height of the two lines is to be expected, just as a result of chance. But if the lines cross the thin gray horizontal line at the top, which reflects the conventional criterion for statistical significance ($p = 0.05$), we should conclude that a real change has taken place.

As we can see from the figure, the solid and dashed lines don't cross the gray horizontal line at the top at all. For nearly every point in time, they don't come especially close. There is only one cluster of cut-points, around 1945, that even approaches statistical significance. What the figure tells us, then, is that the deadliness of the wars prior to 1945 and the deadliness of the wars after 1945 may have differed somewhat, but not so much that the difference can't be attributed to chance.

These tests are extremely useful, in that they'll flag any difference at all in the distribution of war intensity over time. For power-law-distributed data like war intensity, however, most of the action is in the thick tails. That's why it's also worth examining the best-fit lines through the data in a log-log plot, as Cederman, Warren, and Sornette do. Those lines tell us about the escalatory potential of wars—the probability that a small war will snowball into a very large one. That escalatory potential is interesting both in its own right and because it's the main determinant of average war size, so comparing the slopes of these lines is a handy way to tell whether there has been a change in the average deadliness of war.

If we take a look at the long tails of the pre-1945 and post-1945 distributions of war intensity (Figure 5.4), our conclusion about the absence of change is confirmed. The two lines—gray for the period from 1815 to 1945, red for the post–World War II period—are almost exactly parallel. Their slope coefficients scarcely differ (1.67 prior to 1945 versus 1.59 afterward). In fact, to the extent that the intensity of war in these two periods does differ, the results are bad news for the decline-of-war theorists: the slightly shallower slope of the post-1945 tail distribution indicates that wars got *more* deadly, not less, after World War II.

Is the difference in the slopes of the two lines big enough to warrant the conclusion that such a change is real and not just statistical noise?

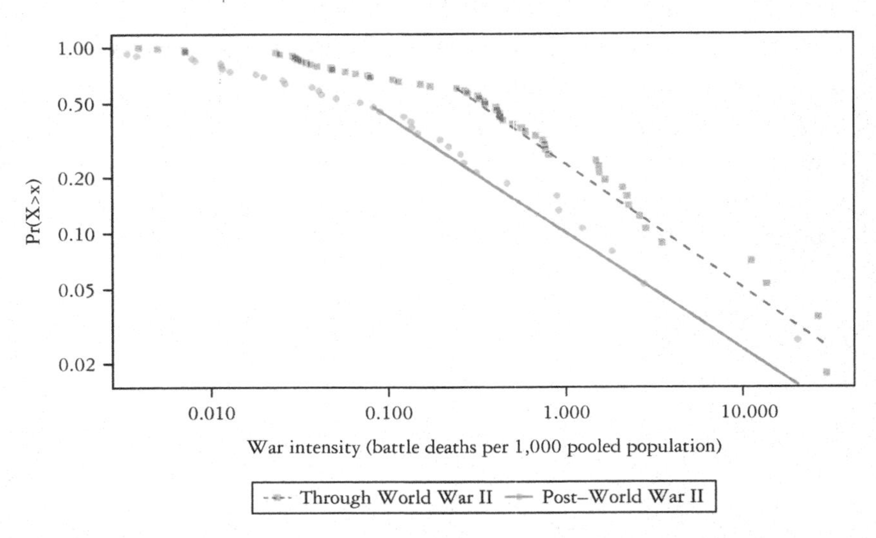

FIGURE 5.4 The long tails of the distributions of war intensity before and after 1945. Shallower slopes correspond to periods with deadlier wars.

We can't answer that question just by looking at the lines themselves. We need to take into account the uncertainty of our estimates of the slopes of those lines before we can say with any confidence whether they're really different.

I measure that uncertainty by using a clever statistical technique called *bootstrapping*. The idea that motivates bootstrapping is that, because outcomes depend to some extent on chance, all of your observations can stand in for other observations that might have happened but didn't. The German invasion of France in 1940 could have failed, and the Germans could have been defeated with about as much bloodshed as (say) the Russo-Finnish War. Or the second Balkan War could have spiraled out of control and produced a war on the scale of World War I. Or President Truman could have listened to the National Security Council's recommendation against crossing the thirty-eighth parallel during the Korean War rather than siding with the Joint Chiefs, thereby avoiding Chinese intervention and bringing the conflict to a close with far less loss of life. And so on.

That insight—that the histories that did happen can give us some traction on the ones that didn't—is what makes bootstrapping possible. The technique is brilliant in its simplicity: We simply resample

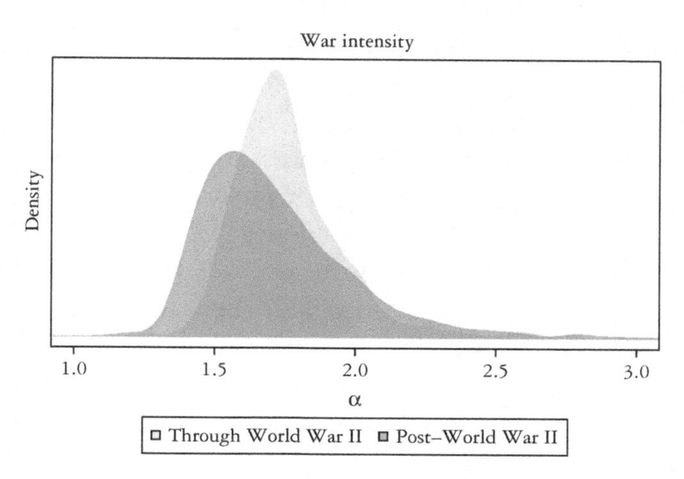

FIGURE 5.5 Distribution of bootstrapped slope coefficients of the long tails of the distributions of war intensity before and after 1945. Distributions that overlap quite a bit, as these do, indicate that the observed difference in the escalatory potential of wars is very plausibly due to chance.

observations from our existing sample, again and again, and use them as estimates of how history might have turned out. We then measure the variation in α across all of these alternative histories to get a sense of how much the slopes of the lines in Figure 5.4 might have varied, purely by chance. Comparing the distributions of simulated slope coefficients allows us to assess the claim that the difference between the two slopes is systematic, rather than just the result of chance variation.

In Figure 5.5, I do just that. The two distributions in the figure are distributions of plausible slope coefficent values, calculated from a thousand alternative histories each. As we can see, there's a lot of variation in the plausible values of these coefficients once chance has been taken into account. In 95% of these alternative histories, the slope coefficient in the 1815–1945 period fell between 1.46 and 2.42, and the slope coefficient in the postwar era fell between 1.37 and 2.53. That much chance variation *within* periods means that the difference *across* periods, a relatively modest 0.08, could very easily have come about as a result of chance variation—Hitler's incredibly improbable victory over France, Truman's decision to listen to the Joint Chiefs—than as a result of any meaningful, systematic difference in the process by which small wars snowball into big ones. A formal test[10] puts a concrete number on that probability: Differences as great as the observed one or greater would occur by chance about 90% of the time. Standard practice in statistics is to hold out for a much lower number, typically 5% or less, before we conclude that a change in the deadliness of war is big enough that it can't just be attributed to random chance. So taking into account the role of chance, we can't conclude that there's been any significant change in the intensity of warfare over the last two hundred years.

Although it may not be immediately obvious, this is exactly the test that Professor Pinker and his colleague, Professor Michael Spagat, point to as being the gold standard for evaluating the claim that there has been a change in the deadliness of war:

> [W]e are perfectly aware that a stretch of time without a big war does not imply that a big war cannot happen. *The issue is whether the parameters of the processes generating new wars and determining their magnitudes have changed since 1945.* This is the era that historians have called "the Long Peace," in which wars between great powers and wars between developed states, common throughout recorded history, essentially disappeared. In the familiar analogy of drawing balls from urns, the idea is not... whether drawing a series of balls with low numbers is taken to suggest that the urn contains no balls with very high numbers; it is whether there is reason to suspect that the urn has been tampered with so as to change the number of balls with numbers of various sizes. (Spagat and Pinker, 2016, 44, emphasis added)

The key parameter of the process that generates new wars and determining their magnitudes is α, the slope coefficient that describes a power-law relationship. These figures show us, quite simply, that the data don't support the claim that α has changed since 1815. In other words, there is no evidence to support the claim that the urn has been tampered with. If the urn *has* been tampered with, the tampering most likely involved adding balls with very high numbers rather than subtracting them. But by far the most reasonable answer, given the data, is that no tampering has taken place.

It's hard to understand just how grim this conclusion is without a few concrete examples based on the post-1945 line in Figure 5.4. What that line tells us is that once an armed conflict passes the thousand-battle-death threshold, there's about a 50/50 chance that it will end up being at least as deadly as the 1990 Iraq War. There's about a 25% chance that its intensity will surpass that of the Yom Kippur War, which demonstrated the lethality of state-of-the-art military machinery. As we get further into the thick tail of the distribution, things get worse pretty quickly. There's about a 10% chance that a war will end up being deadlier than the Korean War, which was impressively lethal despite its brevity. There's a 2% chance—about the probability of drawing three of a kind in a five-card poker game—that such a war would end up being about as deadly as World War I. And there's about a 1% chance that its intensity would surpass that of any international war fought in the last two centuries.

As bad as all of that sounds, the mathematical reality of power-law distributions actually has an even worse implication in store for us. The average of a power law distribution—here, the average size of a war—is driven overwhelmingly by the thickness of the thick tail. When the α coefficient that I mentioned above is greater than 2, the mean of the distribution is simply $x_{min}\frac{\alpha-1}{\alpha-2}$, where x_{min} is the smallest value of the variable that conforms to a power-law distribution (typically, only the larger observations do). When $\alpha \leq 2$, the mean of a power-law distribution becomes infinite.[11] What that means is that, with very low probability—almost infinitesimally small—an observation drawn from a power-law distribution with an α coefficient of 2 or below could reach infinity. And because infinity is a possibility, the sum of all possible outcomes is infinity, and the mean of the distribution is infinite.

In practice, of course, it's not possible for the number of war deaths to be infinite. War deaths cannot exceed the combined populations of the combatants, so there's an upper bound on the intensity of war—a thousand battle deaths per thousand pooled population—that isn't reflected in the power-law formula. The power-law line *has* to curve downward or just hit its logical limit, eventually. But neither of the lines in Figure 5.4 shows any sign of doing so, and there's no way to know how bad wars have to get before they do.[12]

With that uncertainty in mind, I think the best way to interpret these results is to say that there is no obvious upper limit to how big wars can get, short of complete annihilation. That's as true today as it was two hundred years ago. If the power-law relationship holds all the way up to the logical limit, there's about 1 chance in 3,100—significantly better than the odds of drawing four of a kind in a game of five-card poker—that a war that passes the thousand-death threshold would kill everyone involved, right down to the last person.[13] If you recall that the odds of another World War I were about the same as drawing *three* of a kind in a poker game, you'll get a sense of how frighteningly easy it would be to have much, much larger wars than the ones we've already seen.

This news is pretty bleak. It's a relief, of course, to know that things haven't actually gotten worse. But they haven't gotten better, either, and that fact should scare the hell out of you. These results tell us that it's extremely unlikely that the post–World War II period is either more or less deadly than the 130 years that preceded it. And given how deadly the first half of the twentieth century turned out to be, that conclusion is nothing short of horrifying.

Wars Haven't Gotten Less Severe, Either

Again, reasonable people might differ over the best measure to use to gauge the deadliness of war. Even those who agree that war intensity is the best measure of the extent of humanity's propensity for violence in war might still be interested in whether there are changes in severity (raw battle deaths), another commonly used metric of deadliness in the war literature.

As I noted above, I have some reservations about this measure. The raw battle deaths metric doesn't take into account the growth of countries' populations over time—not a trivial consideration even over a period of two centuries. (The United States, for example, had somewhere in the neighborhood of ten million people in 1815, as compared to nearly 325 million today.) It's hard to believe that that difference is totally irrelevant when it comes to war fatalities.

As it turns out, though, it doesn't much matter whether you think intensity or severity is a better measure of the deadliness of war. The change-point detection algorithm (Figure 5.6, top) doesn't register a

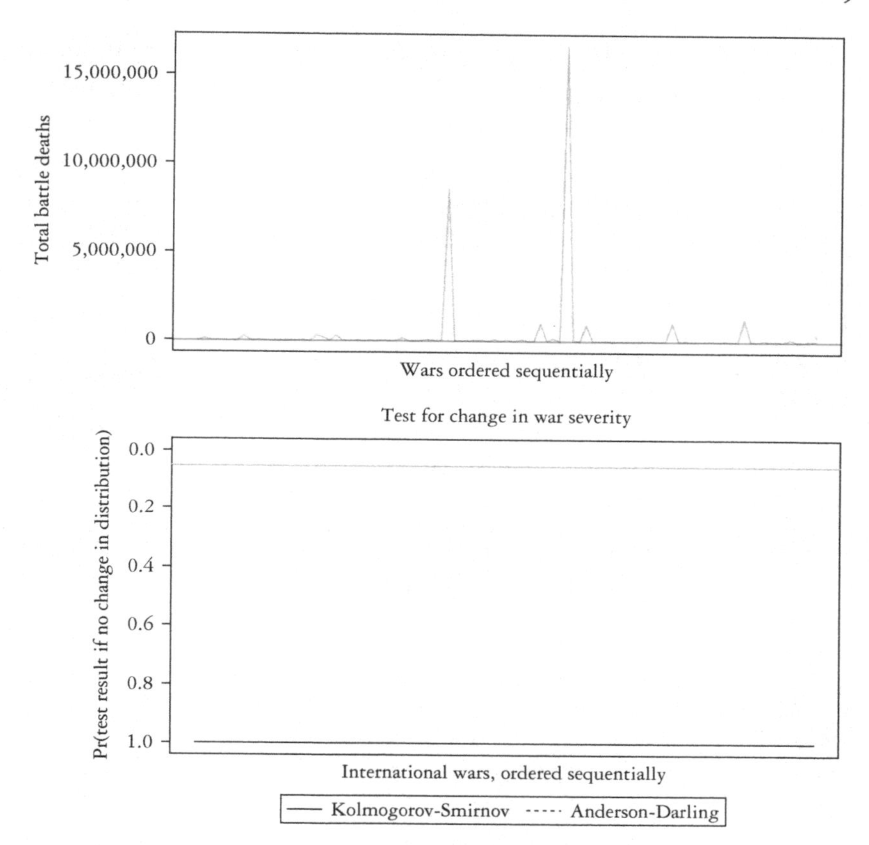

FIGURE 5.6 War severity, with results from the change-point algorithm (top) and Kolmogorov-Smirnov and Anderson-Darling tests (bottom) indicating no change over time.

change in severity over the course of the past two centuries. By this measure, the two World Wars stand out, and no other wars come close. The bottom panel of Figure 5.6 shows the results of the Kolmogorov-Smirnov and Anderson-Darling tests—two flat lines at 1.0, indicating that there's no point between 1815 and the present at which the difference between the earlier and later distributions of war severity are even close to significantly different.

What happens if we look only at the thick tails of the distribution? In short, the data don't support the claim that there's been a decline in either the escalatory propensity of wars or the average deadliness of war over time. A comparison of the slopes of the best-fit lines through the data in the thick tails of the pre- and post-1945 distributions (Figure 5.7, top) show only a modest difference between the two. The post–World War II line is shallower than its counterpart in the pre-1945 era, indicating, if anything, an increase in the deadliness of war. Again, however, the overlap between the bootstrapped distributions of power-law coefficients based on alternative histories (Figure 5.7, bottom) indicates that the difference can very reasonably be attributed to chance: Differences as great as the observed one or greater would occur by chance about 70% of the time.

Here, too, a few concrete examples give a sense of how horrifying these numbers are. If we use severity—raw battle deaths—as our measure of war, there's about a 2% chance that a new war will escalate to roughly the level of the Vietnam War (following the introduction of American ground forces). There's about a 1% chance of seeing a war that's as deadly, in absolute terms, as World War I. And there's *one chance in two hundred* of a war that will produce almost twice as many battle deaths as World War II, the deadliest war in human history. If we keep fighting wars that pass the thousand-battle-death threshold at a rate of roughly one every other year, the probability of seeing such a war in the next century jumps to nearly 40%.

Here, too, the slope coefficient, α, is less than 2, indicating that the mean of the power-law distribution is infinite. For war severity, however, the logical limit of war size is the total population of the planet, just over 7.5 billion people at this writing. And while the probability of a war that big if the power-law relationship holds to the limit isn't huge—roughly three chances in ten thousand—it's not really comfortably small, either, especially if we keep rolling the dice over a long enough period of time.

All told, the picture here is every bit as bleak as it was when we used intensity as a measure of the deadliness of war. The trajectory traced by the wars we have had since 1945 indicates that we live in a period in which wars are every bit as prone to escalation as they used to be. The specific historical conditions that produced war in 1914 and 1939 will never be reproduced, but the dynamics that turned them into World Wars still operate today.

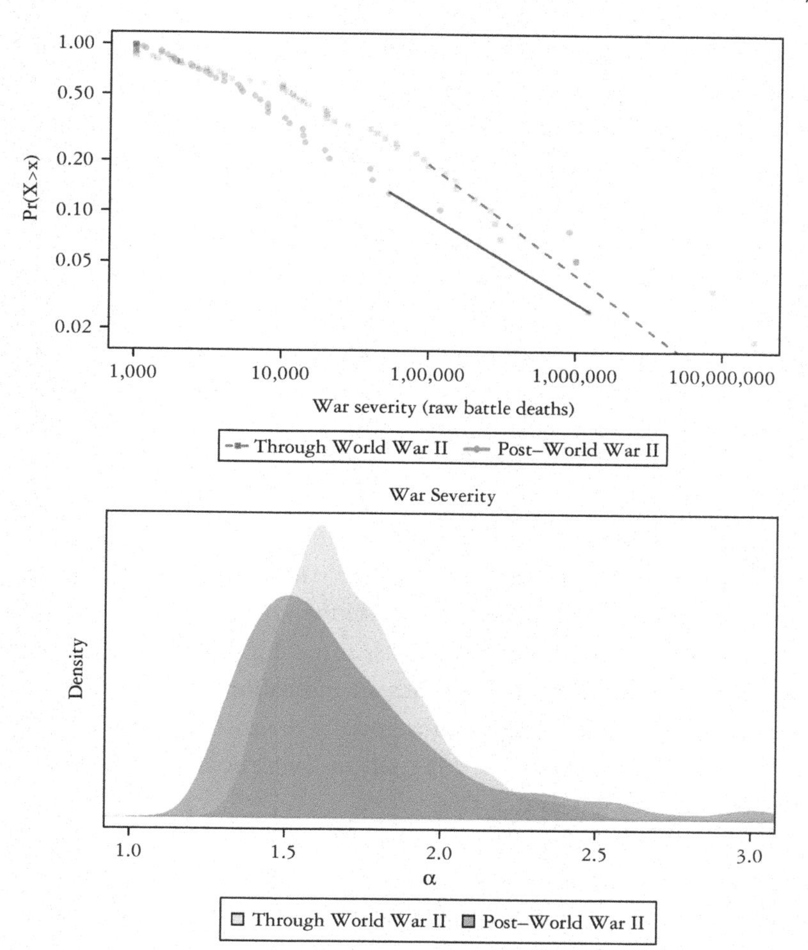

FIGURE 5.7 The long tails (top) and the distributions of bootstrapped slope coefficients (bottom) for raw battle deaths, before and after World War II. Distributions that overlap quite a bit, as these do, indicate that the observed difference in the escalatory potential of wars is very plausibly due to chance.

War Prevalence *Has* Changed...

What about yet another measure of the deadliness of war—war prevalence, or the number of battle deaths divided by world population? I mentioned above that I have real qualms about this measure's usefulness as a way to get at the deadliness of war. One thing I've learned, though, both in paper presentations and in one-on-one discussions, is that if you think this measure is the one that matters, I won't have changed your mind by now, and I probably never will. Some people—some very smart people—see war prevalence as the most important measure of the deadliness of war. As I noted earlier, that's typically because they want to know something subtly different than what I've set out to uncover here: They want to understand how bad war is as a worldwide public health problem, whereas I want to understand the nature of war as a form of organized human behavior. *De gustibus non est disputandum:* There's no point arguing about which one we should care about. (It also doesn't matter because we're screwed either way, but I'll get to that in a few minutes.)

First, the good news. When we run the same tests on these data, we do find something different. The change-point algorithm continues to show no change. But when we turn to the Kolmogorov-Smirnov and Anderson-Darling tests, some cut points *are* statistically significant, albeit just barely. That's an indication that the earlier and later distributions differ more substantially than we would expect them to due to chance. The tests don't tell us *how* the distributions differ, exactly, but they tell us that they differ.

So let's dig a little deeper. The easiest way to figure out how the distributions differ is simply to plot them side by side. Figure 5.9 is a visualization of *empirical cumulative distributions,* which show you the fraction of wars that are equally or less deadly than a war of size x for all values along the x-axis. If you follow the lines from left to right, the post-1945 line rises more quickly than the pre–World War II line. What that means is that a larger fraction of the wars in the post-1945 period fall at the low end of the distribution than had been the case prior to World War II. Put more simply, there were more small wars—"small," again, as a percentage of world population—after World War II than before.

But the Danger of Escalation Hasn't Changed

That said, the most important question from the point of view of the decline-of-war debate is whether the escalatory dynamics that allow

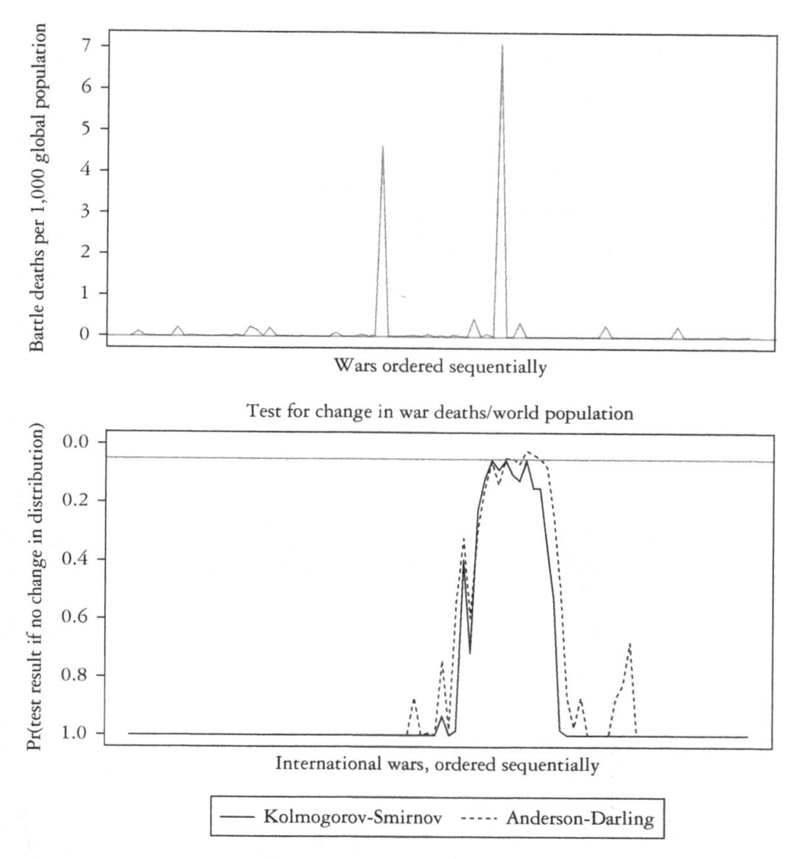

FIGURE 5.8 War prevalance, or battle deaths per thousand world population. Results from the change-point algorithm (top) indicate no change over time, while results from the Kolmogorov-Smirnov and Anderson-Darling tests (bottom) indicate a significant change around the time of World War II.

smaller wars to snowball into larger ones have changed since World War II. The smallest wars have gotten smaller, at least by this measure—but do they still retain the capacity to spiral into a massive, lethal World War III?

Unfortunately, Figure 5.10 suggests fairly strongly that they do. The slope coefficients for the pre- and post-1945 distributions (top) are nearly identical—1.54 and 1.58, respectively—and the distributions of bootstrapped slope coefficients (bottom) indicate that the difference

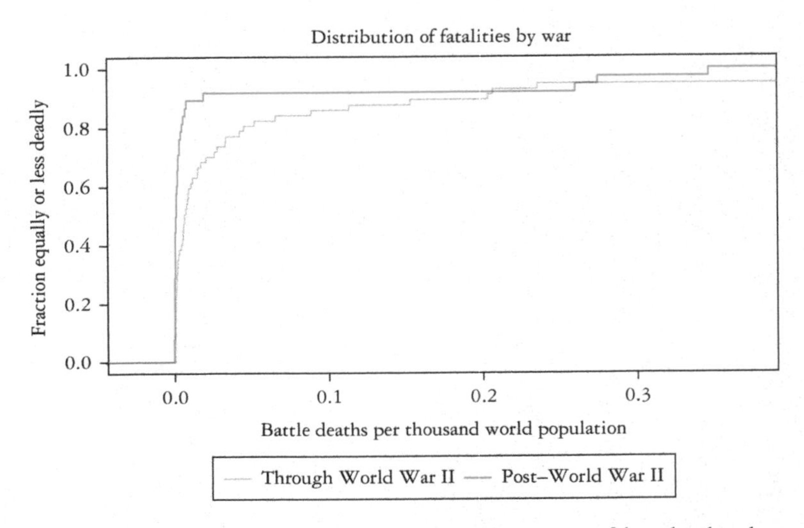

FIGURE 5.9 The empirical cumulative distributions of battle deaths per thousand world population in the years up to and including World War II and the post–World War II era.

between the two of them is small enough that it is very plausibly the result of random noise. Indeed, the two distributions are virtually identical.[14] Given the similarity of the two coefficient estimates and the degree of uncertainty surrounding them, we'd expect to see differences as large as the observed one or larger more than 85% of the time by chance—so we really can't conclude that there's been a change in the escalatory potential of war.

And by this measure, the potential of war to escalate remains nothing short of horrifying. Based on the post-1945 data, there's about 1 chance in 350 that battle deaths from the next war will exceed 1% of the world population. Those are long odds, but they're by no means impossibly long: If we continue to fight roughly fifty wars per century, the probability of seeing a war that large or larger in the next hundred years is about 13%. The probability of seeing a war that kills more than 5% of the world's population in battle over the same time period is a hair over 5%.

And That's What Really Matters

These numbers might seem pretty jarring, given that I started off this section by noting a decline in war deaths as a percentage of

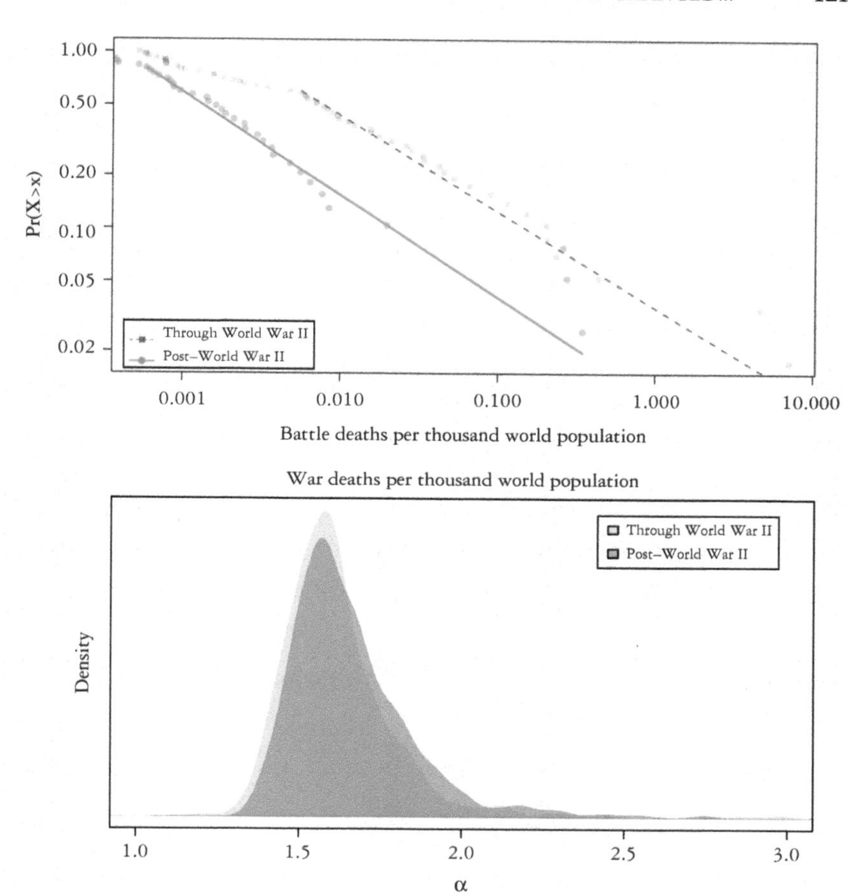

FIGURE 5.10 The long tails (top) and the distributions of bootstrapped slope coefficients (bottom) for battle deaths per thousand world population, before and after World War II. Distributions that overlap quite a bit, as these do, indicate that the observed difference in the escalatory potential of wars is very plausibly due to chance.

world population. The bottom line, I think, is that it has gone down numerically, but not theoretically. As a result, the difference doesn't matter.

Let me unpack that a little. What I'm calling war prevalence is simply war severity (raw battle deaths) divided by world population. World population has gone up a *lot* in the past two hundred years, from somewhere in the neighborhood of a billion people in the early 1800s to around 7.5 billion now. As we know from the previous section, war severity hasn't increased during the same period. So if the numerator hasn't changed and the denominator has increased, we'd expect to see some decrease in observed war prevalence, and we do.

At the same time, the numerator, war severity, still follows a power-law distribution with a mean of infinity. And infinity divided by any finite number, even 7.5 billion, is still infinity. So when you take the theoretical properties of the distribution into account rather than just the data we've observed in the past two hundred years, you realize that the average war size hasn't really changed. In practical terms, a growing world population does mean a shrinking probability of a war that would result in the total annihilation of the human race. But the escalatory potential of war is so spectacular that that probability has only changed by about five one-hundredths of 1% over the last two hundred years, a difference that could very easily just be the result of statistical noise.

Caveat: Militaries Don't Keep Up With Population Growth

There's a challenging wrinkle when it comes to using per-capita measures of the deadliness of war to produce inferences about human behavior that's worth noting here. Recall the arguments by Professors Oka et al. and Falk and Hildebolt that population growth increases security and reduces the need to have a large military, and that those (relatively) smaller militaries account for lower death rates from war in larger societies. Both sets of authors demonstrate that these relationships hold across very wide ranges of human (and even chimpanzee) society size. Historian Azar Gat (2013, 151) makes the same argument regarding the modern state. Such a trend could easily explain the apparent decline in the prevalence of war among small and medium-sized countries over time. That possibility is worth exploring.

The first part of the argument isn't hard to assess and, you might be glad to hear at this point, doesn't even require figures. World population more than doubled from 1950 to 2000, from about 2.5 billion people to about six billion. At the same time, the average size of a country's militarized forces worldwide actually *decreased*—from about 239,500 in 1950 to 187,700 in 1970 to 174,000 in 1990. It took a considerable drop in the decade after the Cold War ended, to just under 120,000.[15]

So, the first part of the argument is certainly plausible: On average, military forces aren't growing at anywhere near the rate that the global population is. In fact, in recent years they've even been shrinking.

We can say more than that, though. If larger societies lead to smaller militaries and thereby to smaller wars, we should see a negative relationship between the populations of states that fight and the intensity of the wars that they fight—and that relationship should hold more or less steady over time. Population, not the passage of time, should be associated with decreasing war intensity. If, on the other hand, people are becoming less warlike over time *and* larger societies produce relatively smaller wars, we should see decreasing war intensity as a function of population size across time periods as well as within them.

Figure 5.11 illustrates the distinction between the two scenarios. In each, there are three separate periods: the nineteenth century, 1901–1945, and the post-1945 era. If larger population size leads to less intense wars, we should see a negative relationship between the size of the populations of the combatants in a war and the intensity of that war. If there is no decrease in the deadliness of wars over time, that relationship shouldn't vary too much across time periods, because time has no effect on war intensity *independent of its effect on population size.* That's what we can see in the left-hand plot.

In the right-hand plot, we see what should happen if the correlation between population size and war size is spurious—that is, if the passage

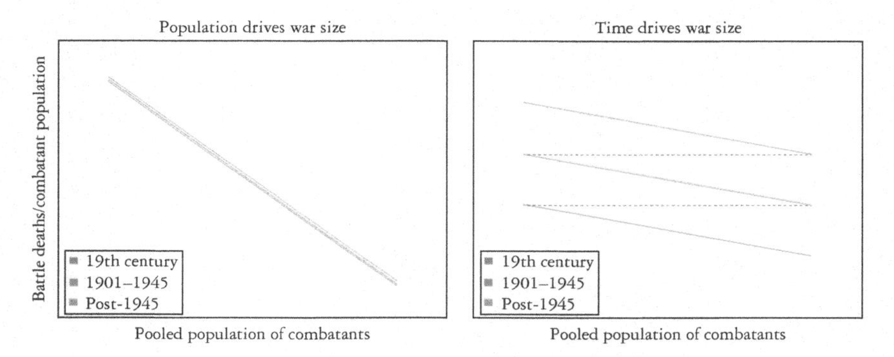

FIGURE 5.11 Ideal illustrations of the relationship between population and war intensity if intensity decreases with population size (left) and if it decreases over time (right).

of time produces both larger states and less deadly wars. There would still be a modest downward slope to the lines, just because countries later in the period are both bigger and more peaceful than they are earlier in the period. But there should be a very noticeable difference across periods as well. In principle (that is, in a world with much less noisy data than the world we actually live in), we might even see continuity between the end of one period and the beginning of the next, as indicated by the dashed gray lines. Even if we don't get such neat, orderly data, though, we should still be able to make out a steady downward progression in war intensity across periods.

When we take a look at the data (Figure 5.12), we can see pretty clearly that no such progression is in evidence. The individual years that I use as data points are the midpoints of the wars that they represent.[16] The lines show the central trend in the data.[17] The overall trend for the post-1945 period looks remarkably like the overall trend in the nineteenth century: while there's a very slight change in slope, there clearly isn't any downward trend across periods. The line representing the 1901–1945 period has a markedly different slope, mostly because it contains two huge outliers in the form of the two World Wars. Were they removed, the line would look very similar to the other two.

This plot gives us some idea of what's going on. If countries with bigger populations fight smaller wars, relatively speaking, we should see a negative relationship between population and war intensity. We do, and it's both strong and robust. If the spread of humanitarian norms were driving both the growth in population and the decline in war intensity, we should see a steady downward trend across periods as well as within them. We don't. For that matter, we should also see decreases in war severity and war intensity across time, and as I demonstrated earlier in this chapter, we don't. So if you want to claim that increased pacifism led to a lower incidence of war deaths worldwide, these results leave you with a lot of explaining to do.[18]

Interpretation

I've just conveyed a fairly detailed and sometimes not-very-intuitive set of findings. What should we make of them? What coherent story about war escalation and deadliness emerges from these data-based snapshots?

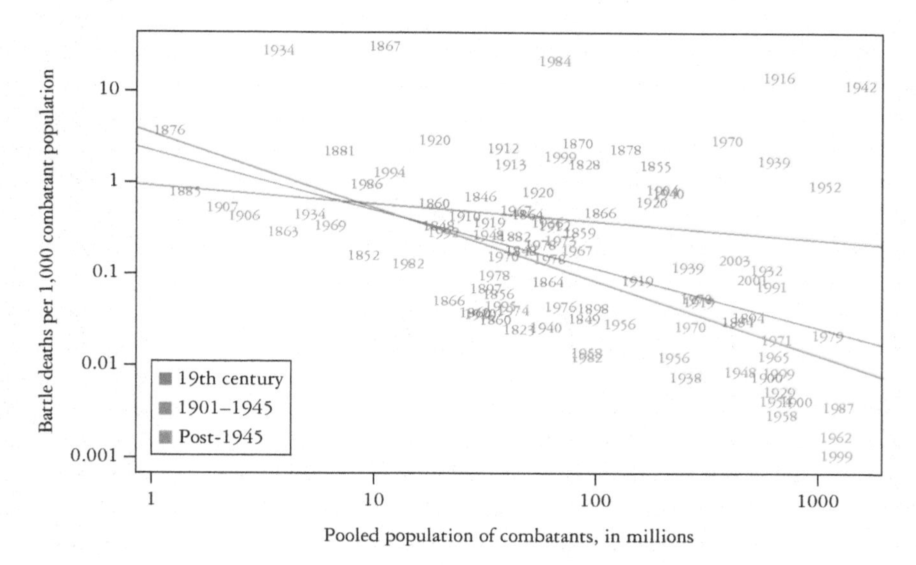

FIGURE 5.12 The relationship between population and war intensity across three periods spanning the nineteenth, twentieth, and twenty-first centuries.

Let's start with the fact that when we examine war intensity and war severity, we see patterns of escalation consistent with a power-law distribution. That fact tells us two things. First, the distribution of war outcomes tells us that while most wars are small, some are unbelievably huge. They are unbelievably huge *even by the standards of other unbelievably huge things.* The low value of α, relative to its value for other phenomena, tells us that wars escalate more quickly, and reach more astounding sizes, than just about anything else found in nature. In their own study, Professors Cirillo and Taleb make exactly this point:

> [W]ar turned out to be the mother of fat tails, far worse than the popular 80/20 rule: there are few phenomena such as fluid turbulence or thermal spikes on the surface of the sun that can rival the fat-tailedness of violence. (Cirillo and Taleb, 2016*b*, 2)

Second, the fact that both intensity and severity are not just thick-tailed but also fit a power-law distribution well is especially ominous. That fact means that the outcomes of wars are consistent with a process in which the only difference between a small war and a very large one is random chance.

Even people who can manage to wrap their heads around the incredible escalatory potential of war often find this point hard to believe. Don't human beings have control over the outcomes of wars? Can't we just put a stop to them if they're getting too bloody? There's a short answer to this objection and a longer, more detailed one. The short answer is that, yes, we certainly can put a stop to a war that we're fighting—if we're willing to lose. But if we were willing to lose, we probably wouldn't be fighting the war in the first place, would we?

The longer answer is that it's not inconsistent both to believe that we can stop wars when we want to *and* that chance plays a huge role in the outcome. Let me offer a simple analogy to give you a sense of what I mean. Imagine that the logic of combat in a war, in the abstract, resembles a game in which you and I are facing off against each other on the fifty-yard line of an American football field. Each of us wants to conquer the whole football field. It's costly to do so, but we're willing to gamble a fair amount of money to try. So on every round, we flip a coin. If the coin comes up heads, I've won that round, and we move one yard toward your end zone. If the coin comes up tails, you've won that round, and we move one yard back toward my end zone. Once we've moved, each of us burns $1. Then we try again, and again, and again, each of us burning a dollar after every move. The game ends when one of us is backed up so far (let's say, to the ten-yard line) that person surrenders because the dwindling odds of winning no longer justify burning any more dollar bills.[19] Chance, and chance alone, determines the outcomes of the coin flips, but human calculation drives the decision to surrender.

While this abstract game obviously doesn't capture all the nuances of actual combat, it does highlight some essential features. The coin flip stands in for the outcomes of individual battles, which owe more to chance than any of the participants would like. Indeed, no less a commentator than Carl von Clausewitz (1976 [1832], 85) argued that "[n]o other human activity is so continuously or universally bound up with chance." The dollar bills represent the costs of those outcomes to the participants. The decision to continue or capitulate neatly captures the main calculation that leaders have to make: does the probability of eventual victory justify the likely cost of fighting to achieve it?

The football-field game also helps us get a handle on the process by which wars can go on for a lot longer, and be a lot more expensive, than you might expect. The coin flips that determine where you are on the field tend to balance out over time. That tendency, known as the law of averages, doesn't mean that flipping heads makes tails more likely in subsequent tosses; the coin flips are independent of one another, meaning that tails is no more or less likely on any toss than it would be on any other. It just means that, as the number of tosses increases, the overall balance of heads and tails tends toward 50/50. As a result, you should expect to spend a lot of time—far more than unaided intuition would suggest—moving around the middle of the field without either side gaining much of an advantage.

How long could this game go on? Well, you could theoretically get to your opponent's ten-yard line by burning as little as $40, but that would virtually never happen in practice—it'd require flipping heads forty times in a row. You might expect the average cost of a game to be five or six times that great, maybe $200 or $250—but you'd soon find that fewer than 5% of the games you play cost you that little. Maybe you'd expect the average cost to be ten times as great, about $400? Maybe, but not likely—that would happen less than 15% of the time. No, if you start playing this game at the fifty-yard line, you should expect *on average* to have burned $1,600 before the game ends, one way or the other. (Even if you make it to your opponent's twenty, only ten yards from your goal, you should expect to burn $700 more before one of you yields.) And that's just an average game—you could easily burn $7,500 or more just to end up forty yards from where you started.

These shockingly expensive contests, in which the combatants spend most of their time around midfield, capturing and recapturing the same small stretches of territory, bring to mind the sort of trench warfare that took place throughout World War I and the Iran-Iraq War. In fact, if you simulate this game lots of times, you get a power-law distribution of dollars burned, one that looks a lot like the distribution of war deaths (Figure 5.13). That's not ironclad proof that this is the mechanism by which wars snowball, of course.[20] But because it produces similar outcomes, and because it neatly captures the back-and-forth dynamics of real wars, it strikes me as being a particularly intuitive way of understanding how the outcomes of wars could be consistent with a

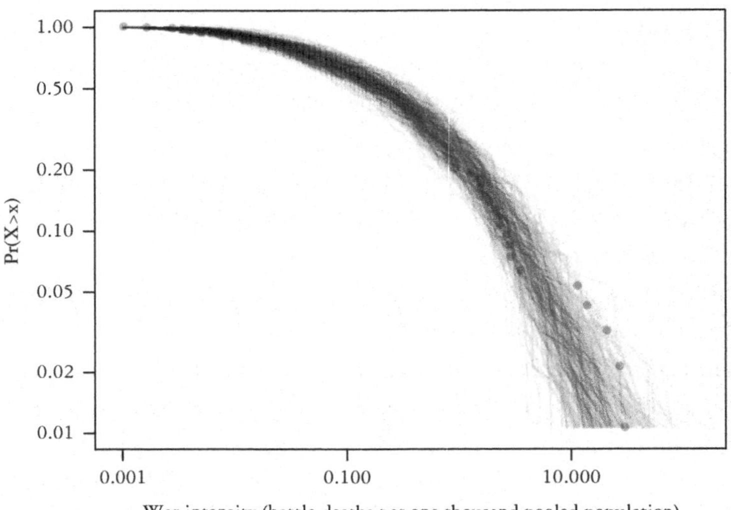

FIGURE 5.13 The distribution of war intensities (red dots) compared to the outcomes of 250 sets of 95 simulated wars each (gray lines), generated via the simple football-field model described in the text and calibrated to match the scale of the actual war outcomes.

process in which the only difference between a small war and a very large one is random chance.

Now, armed with that analogy, how should we think about the fact that there's been no detectable change in war intensity or severity since 1815? Ideally, we should start by pouring ourselves a very stiff drink, because the implications are terrifying. The burn-the-dollar game, like war, produces lots of modestly costly contests and a few that are horrendously expensive. The same calculations and decisions that produce the ordinary ones also produce the catastrophic bloodlettings. People don't have to be depraved monsters to fight wars that kill millions. Nor do they have to be irrational. Thanks to random chance, it's possible for rational, enlightened leaders to arrive at an outcome that nobody would have chosen via a series of steps that everyone agreed on.

And it gets better. If chance is the main driver of escalation, it follows that we can never know when another incredibly lethal war is just around the corner. You might be skeptical of this claim because we haven't seen too many wars lately that have featured out-of-control escalation. If we leave aside the bloody Iran-Iraq War (which, unfortunately, most discussions of this topic do), that's more or less true. But most wars *don't* spiral out of control. That doesn't mean that they couldn't have, and it doesn't mean that the next one won't.

For that reason, we shouldn't be lulled into complacency by talk of a post-1945 "Long Peace." What these results tell us is that nothing about the way that war is fought today stands in the way of a horrific bloodletting, even a cataclysm on the scale of (or exceeding) the two World Wars. Leaders of countries cannot enter into war safe in the knowledge that the world has become a safer, nicer place. It has not. The escalatory dynamics that produced mindboggling carnage in the Paraguayan War, the Chaco War, the Iran-Iraq War, and the two World Wars are every bit as operational today as they were then, and the potential carnage every bit as unimaginable.

What about the finding that war prevalence—battle deaths divided by world population—does seem to have changed over time? I'm not inclined to make much of it, to be honest. I wish I were. It doesn't mean that wars are killing fewer people in absolute terms, or even relative to the combatants' populations, which I suspect are the measures that matter most to people who are actually involved in wars. It probably *does* mean that population will continue to grow more quickly than battle deaths, as long as wars remain small or even medium-sized. That's not nothing, of course. But all bets are off when it comes to big wars. The more important finding by far is that, even if we measure deadliness as battle deaths divided by world population, there's been no change in the propensity of small wars to snowball into enormous massacres. Presumably, if humanitarianism or peacekeeping were making wars less fatal, we should see a decrease in the slope of the long tail of the post–World War II distribution in Figure 5.10. We don't.

That conclusion would seem to fly in the face of quite a lot of data from recent years that indicates that, in the big picture, the risk of death from war is quite low. Cardiovascular disease and cancer are more deadly, of course, but in most of these tallies even fire, drowning, and suicide kill far more people than war. Those statistics aren't wrong. But they are very misleading, because war's lethality is so incredibly, unpredictably volatile. Cancer will probably kill roughly the same percentage of the population next year as it did this year, give or take a little. That's true of most causes of death—cardiovascular disease, diabetes, malaria, and so on. But it's not true of war. In the early 1930s, the global death toll from war was also low, and it remained low until World War II killed about 3% of the world's population. No other cause of human mortality exhibits war's ability to go from minor nuisance to global catastrophe in the blink of an eye.

Conclusion

When I sat down to write this conclusion, I briefly considered typing, "We're all going to die," and leaving it at that. I chose to write more, not because that conclusion is too alarmist, but because it's not specific enough.

Before I started this project, I thought I understood what it meant for war intensity to follow a power-law distribution. I didn't. I understood that, when you rank wars by severity and plot them on a log-log axis, you'll get a straight line, and I understood the implication that very large wars could theoretically happen. What I didn't understand was the dynamics that produce power-law distributions. I didn't understand just how deadly those big wars could be. I didn't understand how easy it would be for an ordinary war to spiral into an incomprehensibly lethal one. And I didn't understand that, very possibly, the only thing preventing that from occurring was chance.

If the parameters that govern the mechanism by which wars escalate hasn't changed—and there's no evidence to indicate that they have—it's not at all unlikely that another war that would surpass the two World Wars in lethality will happen in your lifetime. And if it is bigger than the two World Wars, it could easily be a *lot* bigger. That's a possibility that no sane citizen, or world leader, can afford to ignore.

6

Are the Causes of International Conflict Becoming Less Potent?

FINALLY, WE CAN TURN to a question that's central to most decline-of-war arguments but has yet to be addressed head-on: Are the causes of war becoming less potent? Does the same stimulus today produce a less belligerent response from potential it did twenty or fifty or a hundred years ago?

As we'll soon see, the most accurate and succinct answer to this question is, "Yes and no." Some causes of war have become more potent over time. Others have become less potent. Many follow no obvious trend one way or the other, even over the course of decades. The results from this chapter indicate that the potency of the causes of war varies dramatically even over short periods of time but rarely trends in one direction or another for very long. While that might be alarming news to practitioners, whose statistical models often don't account for such fluctuations, it ends up telling us very little in the aggregate—except that, *contra* the expectations of the decline-of-war school, no overall downward trend is evident.

The Potency of the Causes of War

I realized that the potency of the causes of war was an important question to explore when I was discussing this book with Professor Andrew Moravcsik at a political science conference a few years ago. Andy and I co-taught an introductory graduate seminar for a few years, and he's one of those people who can formulate smart objections far more quickly than you can respond to them (well, more quickly than *I* can respond to them, at least). When I told him that I didn't buy the decline-of-war thesis, he responded by rattling off a list of international tensions that had lessened or disappeared since the end of the Cold War. "All of these issues are gone," he said. "How can you say that war isn't on the decline?"

I'm always wary of arguments like these, just because it's easier to notice old things in decline than it is to notice new things on the rise. When my former colleague Robert Putnam (1995) pointed to the decline in membership in bowling leagues as an example of declining social capital, for example, journalist Nicholas Lemann (1996) took him to task for selective attention, pointing out that membership in youth soccer leagues had been increasing at the same time. So as Andy was talking, I started mentally compiling a list of new causes of war that I thought might be supplanting the old. I didn't come up with too many, and I'm pretty sure I didn't end up convincing him. (Of course, the list I came up with *after* our conversation was a much better one.)

As I was mulling over our discussion, something hit me. When I told Andy that I didn't buy the decline-of-war argument, he hadn't responded by arguing that war is in decline. He'd argued, instead, that situations that used to create a real risk of war no longer do so. If people are indeed becoming more peaceful than we were dozens or hundreds of years ago, we shouldn't be as prone to fight today as we were back then *when faced with the same circumstances.* Territorial disputes might still exist, but they'd be more likely either to be settled peacefully or to be ignored than to produce armed conflict. Secessionist claims should now be more soluble short of war than they were in the past. And so on. The majority of the examples that Andy cited had this character to them. The source of conflict—the Spratly Islands, say, or the existence of Israel—hadn't gone away. People were just less likely to fight over them than they had been before.

While that argument hadn't been tested in any of the statistical studies of the decline-of-war thesis that I'd seen, it was articulated well by Simon Kuper (2014) of the *Financial Times* after the recent Russian military intervention in Ukraine. While reasonable people might see this event as evidence that Professor Pinker's optimism about war had been misplaced, Kuper argued that it was actually vindication of his argument. Few Russians or Americans wanted war, he pointed out, and "given declining modern appetites for violence, mass bloodshed looks unlikely." Professor Pinker (2015b) agreed:

> At the time of this writing (early March 2014) international tensions are running high over competing US, European, and Russian interests in a chaotic Ukraine, which has been riven by a coup, ethnic conflict, a Russian military incursion, and possible secessions by Crimea and eastern regions. These are just the kinds of tensions that in the past led to great-power wars with millions of deaths. I predict that because of the changes documented in *Better Angels,* such a war will not take place.

I suspect this prediction was made at least somewhat tongue-in-cheek. As I pointed out in the last chapter and as Professor Pinker surely knows well, wars of that magnitude have always been very rare, so as predictions go, this is a pretty safe one.

The tests in previous chapters don't really test the claim that the causes of war are losing their potency as directly as I'd like.[1] The more I thought about it, the more it struck me as an interesting question. So I set out to answer it.

Measuring Potency

Measuring the willingness of countries to use violence when faced with similar sets of circumstances is a bit tricky. The other measurements in this book are all measurements of a single thing—conflict initiation, war intensity, or what have you. The quantity of interest here is a measurement of the relationship between two things, a stimulus and a response. To get an accurate measurement of the magnitude of that relationship, you need to control for potentially confounding factors, which in turn requires a credible theory of how they're all related. As anyone whose work has been through peer review can tell you, that's a high bar. To add to the complexity of the problem, we need to know not just the magnitude of the relationship between these two quantities but how it changes over time.

A concrete example will help to illustrate these challenges in practice. In 1978, the eminent American sociologist William Julius Wilson

(1978), then a professor at the University of Chicago, published a book with the provocative title *The Declining Significance of Race*. In the book he argued that thanks to state intervention designed to increase racial equality and postwar economic expansion, social class had surpassed race as the primary determinant of upward mobility for African Americans. Let's leave aside the argument about class for a moment and focus on the argument that the importance of race has declined in absolute terms over time. In order to test that argument, we would first need to know what the main outcome is: Wilson himself focuses on earnings. Then we would need individual-level data on both earnings and race over time, which are not too hard to come by (Wilson uses data from the Current Population Survey). In order to capture the relationship between the two, however, we would need to account for other variables that might be correlated with both race and income—education, region, wealth, and so on. It's very difficult to know both that you've chosen the right variables and that you've properly accounted for the influence of those other variables. But even if you've succeeded in doing that, there's another dimension to the problem: You need to figure out whether that relationship has changed over time. Doing all of this is so tricky that, more than thirty years after the original publication of his thesis, Wilson (2011)—now a professor at Harvard—still has quarrels with those who have tried to test it.

If we're going to attempt to capture changes in the potency of the causes of conflict, then—and despite my appreciation of the difficulties involved, I think it's well worth attempting—we need to specify a statistical model of the causes of conflict. Given how much work has been done on this subject in my field already, it would be brazen of me (to say the least!) to claim that I have come up with the One True Model of International Conflict and proceed to use it for these purposes.[2] Moreover, given how many statistical studies of conflict there are out there, it wouldn't be unreasonable for skeptics to wonder whether the Braumoeller Model had been cherry-picked to produce a given result.

To avoid such concerns as much as possible, I've chosen to look at other people's work. More specifically, I re-analyzed statistical studies of international wars or militarized disputes carried out in relevant political science journals over the course of a decade. Each represents a credible model of international conflict by virtue of having passed through peer review at a major professional journal. Each study focuses on the impact of a single variable—democracy, alliances, globalization, trade, etc.—on the probability that a militarized interstate dispute will occur, taking into account the impact of other variables that might confound the relationship. That impact is the quantity of interest. To see whether the causes of war have lost their potency, I will examine trends over time in the magnitude of that impact.[3]

Testing

I asked a team of research assistants to find and replicate studies of militarized interstate disputes and armed conflicts from top political science and international relations journals from the previous ten years.[4] Once those studies had been replicated, I asked them to re-run them using a different statistical model—a generalized additive model, or GAM. While most statistical models used in the study of international conflict assume that the impact of a given cause on the probability of conflict is constant over time—that, for example, joint democracy had the same impact on the probability of conflict in 1820 that it has today—this one allows us to relax that assumption. Once we've done that, we can track changes in the potency of conflict-causing variables over time.[5] I never touched the data myself, and the research assistants were explicitly told to stick to the authors' original formulations.

I've reproduced the results in Figure 6.1. Each one of the lines in this figure charts changes over time in the impact of the conflict-causing variable in question on the probability of armed conflict. A downward-sloping line or curve is indicative of a cause of war that is declining in potency, while an upward-sloping one is indicative of a cause of war that is increasing in potency. It's important to note that despite being grouped together, many of the lines aren't measuring exactly the same thing. In some cases, different lines reflect different measurements of the same concept. In others, differences in direction are driven by different statistical specifications. For that reason, the fact that they occasionally trend in opposite directions isn't as alarming as it might seem.

The results of this comprehensive exercise in replication are pretty unambiguous: there is no systematic downward trend in the potency of the causes of war. Depending on how you count, a bit more than

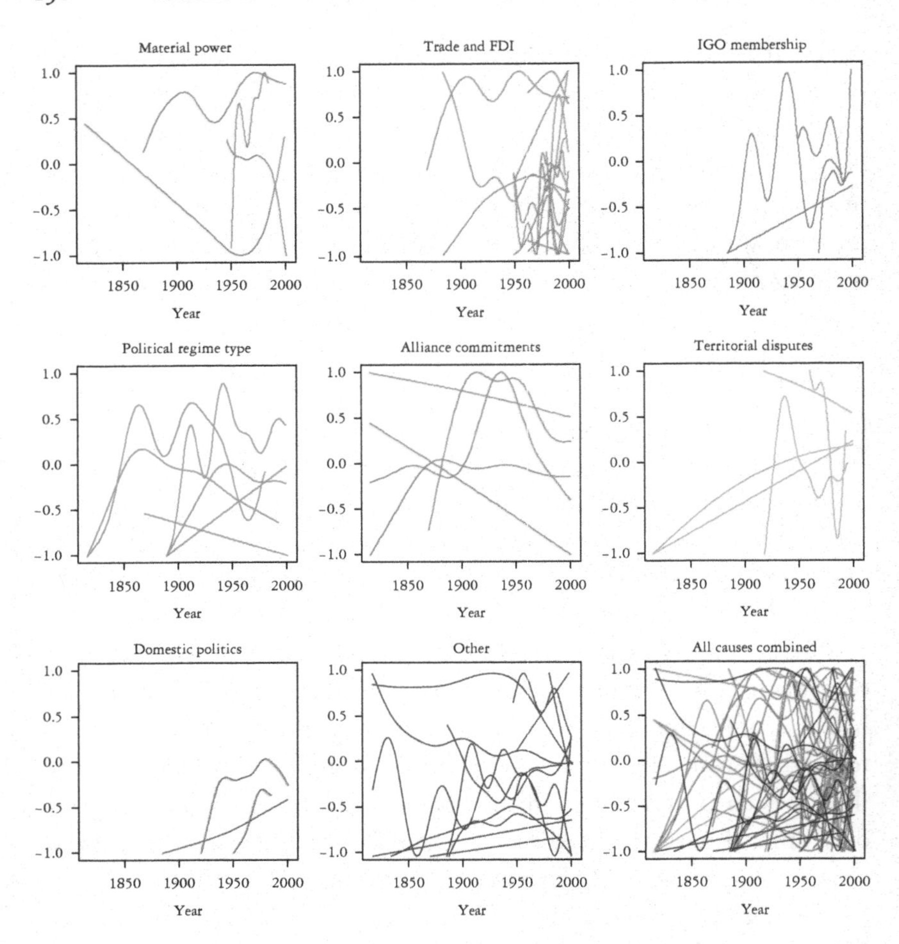

FIGURE 6.1 Changes in the potency of the causes of war, 1815–present. Lines that rise as they move from left to right on the x-axis indicate an increase in the potency of a cause of war (or, equivalently, a decrease in the potency of a cause of peace). Lines that fall as they move from left to right indicate a decrease in the potency of a cause of war. If the decline-of-war school is correct, most lines should be trending downward most of the time.

a quarter of the variables analyzed seem to be becoming less potent over time... but about the same fraction are becoming *more* potent over time. The remainder, nearly half, don't really show any systematic upward or downward trend.

In fact, it's not unfair to say that the results are a bit of a mess. The aggregate pane in Figure 6.1 looks a lot like what you might get if you leave a toddler alone for a while with a pack of crayons and a sheet of paper, and the individual panes aren't much better. That means that the impact of most of these conflict-causing variables fluctuates over time, sometimes pretty dramatically. Some of that volatility is an artifact of the standardized scale—their original scales, which depend on how the variable whose impact they're capturing is measured, vary too widely to allow an aggregate visualization that's both unstandardized and legible. I've reproduced each one in its original scale, with confidence regions, details about the conflict-causing variable in question, and a citation to the original study, in the Appendix so that anyone who's curious can take a look.

Interpretation

People fight over things. The things they fight over change. What these results tell us is that, in very broad terms, the changes more or less balance out in the long run. People who argue that the causes of war are fading away often point to things people used to fight over without fully appreciating the things they're just starting to fight over. As far as the decline-of-war thesis is concerned, while we mostly don't see smooth trends in one direction or another, we can see enough to say that they're not mostly trending downward. It's hard to make meaningful predictions or even to say much more than that about trends in the causes of conflict—they're just too volatile.

Why do we see such remarkable volatility in the impact of conflict-causing variables? Two possibilities come to mind. First, international conflict is a pretty rare event, relatively speaking—at any given time, far more countries are at peace than are at war. Second, different wars have different causes, and some of those causes matter more at some times than at others.

Take, for example, the impact of national capabilities—what we'd colloquially call material power—on conflict in the study by Daina Chiba, Carla Martinez, and William Reed (2014).[6] It's the top curvy green line in the "Material Power" pane of Figure 6.1 and the bottom-center pane in Figure A8 on page 244. The trend captures the impact of national capabilities, or material power, on the probability that a country will become involved in a dispute. The peak of the curve in the early 1900s most likely reflects the fact that a lot of Great Powers got into military clashes in the early 1900s, culminating in World War I. The trough around the time of World War II might seem surprising given that the Great Powers did much the same thing at that point, but more small countries were involved in World War II than in World War I, so material power wasn't as strongly related to militarized disputes in the second World War as it had been in the first. Finally, the upward trend following the 1940s is most likely indicative of the increased tendency of the two superpowers, the United States and the Soviet Union, to get involved in militarized disputes with faraway countries.

Specifics aside, the upshot is that the curvature of these lines can be significantly altered by the clustering together of certain kinds of conflicts, like conflicts that involve powerful countries. Because there aren't many conflicts, the law of large numbers doesn't kick in and these conflicts don't "even out" to a stable average, so the kinds of conflicts that happen to occur in a given year or decade can have a big impact on the potency of this or that cause of war at that time.

It's interesting, to me at least, that these short-term fluctuations often swamp any long-term trends in the data. I hadn't expected to see that outcome. Students of international conflict who use statistical methods will probably be horrified by these results. They should be. Nearly all statistical studies of international conflict use methods that assume that the effect of a given variable is constant over time. That assumption is so wrong, so often, that it should probably be considered incorrect unless proven otherwise. Like many problems, however, this one also represents an opportunity. Understanding fluctuations like these would allow us to bring history into statistical analysis in a way that has only rarely been done by asking *why* the causes of war are more potent in one time period than in another (Wawro and Katznelson, 2014).

One possibility has to do with the fact that these are observational studies rather than experiments—the National Science Foundation frowns on the idea of researchers experimenting on foreign countries to find out what makes them go to war, so we're stuck doing the best we can with observational data. And while some statistical models exist that are designed to account for that fact mathematically,[7] straightforward observational models still dominate the literature. The fact that these are observational models, combined with the fact that different scholars use different control variables and research designs, probably accounts for a lot of the inconsistency across studies.[8]

Even with that caveat in mind, though, it's pretty hard to make the case based on these results that people are becoming systematically less sensitive to the sorts of stimuli that produce armed conflict and war. Observational studies do at least try to control for confounding variables, and it's hard to believe that only the ones that produced upward-trending curves did so poorly. Taken as a whole, these results are very hard to explain if spreading norms of nonviolence were making humanity more peaceful.

Conclusion

If we are becoming better people, in the sense that we are now less likely to fight over the sorts of things that used to cause conflict, we should see a general downward trend in the potency of conflict-causing variables over time. Our ability to tease out such trends is, admittedly, hampered by a surprising degree of short-term variability and by the observational nature of the data. Even so, it's hard to see such trends in the data. While we do see some, we see about an equal number that are trending in the opposite direction. It's very difficult to argue, based on these findings, that increasing human enlightenment or pacifism has led us to turn the other cheek at provocations that would previously have led to conflict.

PART IV
Making Sense of the Data

7

International Order

GRADUATE STUDENTS IN MY seminars have long been familiar with the three questions that structure every discussion that I lead. It's not enough, I tell them, just to tear apart an article that we've just read, as often happens in graduate seminars. Criticism is useful, but it's only useful to the extent that it helps us understand the world. In order to ensure that they don't simply become trained piranhas who rip up anything in their path, I insist that they answer three questions about everything we read:

What's good about it?

What's bad about it?

How do we fix it?

I hope by now you've gotten a sense of my answer to the second question as it relates to the decline-of-war thesis. Chapters 2 and 3 laid out a wide range of reasons for skepticism, both about the argument and about the data offered in support of it. Chapters 4 through 6 took us on an exhaustive tour of the most relevant available data and uncovered little if anything that would give us any confidence that the long arc of history is bending toward peace.

I've also discussed what's good about the decline-of-war thesis. I do think it's fair to say that, on the whole, the modern state makes life more peaceful for its inhabitants than it otherwise would be. I very much agree that peace movements, previously the province of churches and Enlightenment intellectuals, became more popular in the nineteenth century and became true mass movements in the twentieth, even if I don't agree that their spread can be connected to decreases in conflict in anything like a direct, linear fashion. I do think it's reasonable to say that we can make *some* progress in solving the problem of war: We have seen some islands of peace throughout the past two centuries, though I see no evidence that those islands necessarily endure or spread. And I certainly agree that the world has become a less warlike place—at least in terms of international conflict initiation—since the end of the Cold War.

The question, then, becomes, "How do we fix it?" How can we repair and replace the story about Enlightenment humanism and norms of nonviolence so that it will better fit the known facts? What new facts can we look for that would help us to know whether we're on the right track?

The answer, I believe, is to focus on the process by which people create ever-widening spheres of political order. The most familiar example of this process to Western students of history is the development of European political units after the fall of Rome, from small feudal societies to larger kingdoms to the modern territorial state.[1] If we want to understand long-term trends in international violence, it makes sense to examine the next level of aggregation, *international orders,* which are attempts to create political order among states.

In this chapter and the next, I'll lay out the case for viewing historical trends in conflict initiation through the lens of international order. In this chapter, I explain the theoretical reasons for believing that patterns of international order should influence patterns of international conflict. You'll notice that I didn't just write that international order causes peace, because there are good reasons to believe that the relationship between order and conflict is more complicated than that. Under some circumstances, international order might even produce *more* conflict than we would expect to see in its absence. In the next chapter, I offer a brief summary of the history of order and international conflict in the nineteenth, twentieth, and early twenty-first centuries to see how well the argument does in accounting for patterns in the data. It does, I think, quite well—certainly better than a theory that implies an overall decline in violence over time.

To be clear, this is a first cut at an answer. These insights are derived from the history of Great Power politics in the nineteenth and twentieth centuries, which are the periods that I know best. Moreover, even if we accept this revised explanation for why the rate of conflict initiation changes over time (as we should!), that acceptance will be provisional. That's nearly always the case in any science. The goal is not to be perfect, but just to be better, and the next person's goal will be to do better than you did. Egos may suffer a bit in the process, but knowledge wins. In this case, the explanation I offer leaves a number of unanswered questions for future researchers to explore. Is it possible that some of the association between international order and peace can be chalked up to their common origins in a preceding war? Sure it is—the effects are very difficult to disentangle. Is it possible that some international orders are better at producing peace than others, and if so, why? What can earlier, non-Western orders add to our understanding of the relationship between order and conflict? I hope future research will credibly answer some of these thornier questions—especially since I plan to carry out much of that research myself. But those nuances will mostly be of interest to people who study war for a living. My goal here is just to lay the foundation for a different and better understanding of why warfare waxes and wanes over time.

International Order

I should be clear from the onset that when I write "international order" I am not referring to some shadowy world government organized and run by the wealthy and powerful of the Bilderberg Group or the Bohemian Club.[2] Rather, I am alluding to the fact that, from time to time, Great Powers make a concerted effort to influence the calculations of other states with regard to war and peace via the creation of international orders that guide and coordinate their behavior and provide a buffer from the dangerous environment of pure international anarchy.[3] (As I noted in Chapter 3, among political scientists "anarchy" refers not to chaos but to the absence of centralized authority.)

International political orders are, generally speaking, a lot like domestic political orders: They are an attempt to create a set of rules that everyone can live by and in so doing escape the dangers of a lawless, anarchical world. Historical examples of international orders are not hard to find. The Concert of Europe, a.k.a. the Vienna System, followed the Napoleonic Wars, the League of Nations system followed World War I, and two rival international orders, the Western liberal international order and the Soviet Communist order, rose to prominence in the wake of World War II. The timing is not coincidental. Very large and costly wars focus the attention of the international community on creating political structures that might help prevent their recurrence (Ikenberry, 2001).

The decline-of-war theorists do discuss some of the components of international order. Professor Pinker (2011*a*), following Professor Norbert Elias (2000 [1939]), emphasizes the importance of the "civilizing process" in the reduction of violence within states over time, but when at the end of chapter 5 he considers the possibility that this same logic might apply to relations among states—a notion that Elias himself endorsed[4]—he quickly dismisses "world government" as an idea that "lives on mainly among kooks and science fiction fans" (Pinker, 2011*a*, 289). He does suggest that a narrower version of order among states, the Kantian peace among democratic countries (Doyle, 1983*a*, *b*; Russett, 1993; Russett and Oneal, 2001), could very well obtain, but he leaves it at that. Professor Mueller (2007, ch. 2) sketches the path of antiwar sentiment across nine centuries and documents its course over the past century in considerable detail but tends to assume that it will translate more or less unproblematically into a decrease in conflict.

When they do discuss international order, they seem to ascribe little if any importance to it as an explanation. Both scholars, for example, cite historian Sir Michael Howard, who in 1991 wrote that it was "quite possible that war in the sense of major, organized armed conflict between highly developed societies may not recur, and that a stable framework for international order will become firmly established" (Howard, 1991, 176; Pinker, 2011*a*, 254; Mueller, 2007, 3). But they seem to have taken international order to be a description of a peaceful world rather than a set of conditions that might produce such a world, as Professor Howard (2000) himself argued in a more recent work.

In so doing, I believe they inadvertently forfeit the best argument in favor of the decline-of-war thesis. International order is *not* simply synonymous with peace. In fact, war is sometimes used as a means to preserve order. International orders have existed for centuries, but—and this is an important point of near-agreement with the decline-of-war crowd—only in the last two hundred years has the preservation of international peace been among their major goals (Howard, 2000, 2). Some are grounded in the liberal intellectual traditions that arose from the Renaissance; others are not. They influence the behavior of states, not primarily or exclusively by inculcating norms of nonviolence, but instead by resolving some of the issues that lead states to clash and by establishing an environment within which states are less likely to choose war.

The attentive reader might wonder why, if international orders have taken root and spread over the last two hundred years, we don't see a steady decline in the rate of conflict initiation over that same period. For one thing, international order has spread and contracted in fits and starts. We seem especially to appreciate the wisdom of working together to construct peace in the wake of massive, bloody, systemic wars, and the farther we get from those the more we seem to lose our appreciation of the virtues of cooperation. For another, international order actually has very complicated implications for international conflict. Once those implications have been taken into account, the relationship between order and peace isn't as rosy as it might seem.

The first complication to take into account is that while international order may reduce conflict among member states, it can produce *more* conflict elsewhere. To see how this dynamic works, consider an insight from the sociologist Charles Tilly. Writing about the formation of European states, Tilly (1975, 42) penned the succinct and memorable phrase, "[w]ar made the state and the state made war."[5] Warfare was part of the process of the formation of political order within modern European states—an order that was synonymous with a decrease in violence. Once those states had formed, however, they fought wars against one another in order to survive and thrive in the international system. Domestic political order reduces the small- and medium-scale violence that is endemic to lawlessness. In so doing, however, it creates the organization necessary for international warfare on a scale that would be difficult or impossible in its absence.

Tilly's insight about domestic order applies to international order as well. International orders tend to reduce conflict among the states that comprise them. While the decline-of-war theorists do recognize this point,[6] they fail to notice the other side of the coin: While wars make international orders, international orders also make war. In fact, Professor Elias (2000 [1939], 254), whose work Pinker cites extensively, makes precisely this argument:

> [F]irst one castle stands against another, then territory against territory, then state against state, and appearing on the historical horizon today are the first signs of struggles for an integration of regions and masses of people on a still larger scale. We may surmise that with continuing integration even larger units will gradually be assembled under a stable government and internally pacified, and that they in their turn will turn their weapons outward against human aggregates of the same size.[7]

The implication is that when only a single international order exists, we should expect to see more peace overall. Until then, however, competing international orders can and often do fuel conflict throughout the international system, as for example the Western liberal order and the Soviet Communist order did throughout the Cold War. In fact, some of the very characteristics that help international orders maintain peace among their members can increase the odds of conflict among those members and states outside of the order.

To add another complication, the prospects for peace within international orders seem to improve as the scope of those orders narrows. Partial international orders benefit from the fact that they are typically organized and managed by a single major power. International orders that function on a global scale, by contrast, run the risk of reproducing major-power clashes when it comes to the production of order. They can still be better than nothing, and as the next chapter will show I think they have been, at least in the last two hundred years. But they are often not as effective as orders that are smaller in scope can be.

For these reasons, international political orders are a double-edged sword. They can improve the prospects for peace among the countries that subscribe to them by mitigating the causes of war. But they are most effective when they are partial rather than global in scope. Worst of all, they can produce external conflict, especially in the presence of other international orders. There seems to be an inherent trade off when it comes to international order: Partial orders are more effective at producing internal peace, but multiple orders based on conflicting principles of legitimacy make the world a considerably more dangerous place.

What Is Order?

In part because multiple groups of scholars and practitioners have written about international order (see note 3 in this chapter for a brief list), and in part because international order is an intangible thing, there is no one commonly accepted definition of international order. Indeed, the term has been used for virtually anything that could be said to order the international system, from the very practice of dividing territory up into sovereign states in the first place (the "Westphalian order")[8] to patterns of authority among units of virtually any kind (Reus-Smit, 2017, 5) to patterns of activity among states (Bull, 1977, 4) to a set of rules and arrangements created by the most powerful country in the world at a given time (Gilpin, 1981; Ikenberry, 2014*a*).

Despite the variety of definitions on offer, I have yet to encounter one that really captures my own understanding of the phenomenon. So I'll start by offering two very good ones along with an illuminating example, after which I'll build my own definition based in part on the best aspects of each.

Perhaps the most famous definition of international order was offered by Professor Hedley Bull (1977, 8), who described it as "a pattern of activity that sustains the elementary or primary goals of the society of states, or international society." Bull (1977, 13) further argued that a *society of states* exists "when a group of states, conscious of certain common interests and common values, form a society in the sense that they conceive themselves to be bound by a common set of rules in their relations with one another, and share in the working of common institutions."

Dr. Henry Kissinger offers a lucid definition that contrasts usefully with Bull's. He also distinguishes among international, world, and regional orders in a way that I find very helpful:

> World order describes the concept held by a region or civilization about the nature of just arrangements and the distribution of power thought to be applicable to the entire world. An international order is

> the practical application of these concepts to a substantial part of the globe—large enough to effect the global balance of power. (Kissinger, 2014, 9).

Note that Bull's definition of order refers specifically to behavior ("activity") while Kissinger's refers to an idea ("concept"). Here, I am more inclined toward Kissinger's understanding, for two reasons. The first is that a definition of order as behavior runs a serious risk of tautology: Order cannot be said to produce behavior if it is defined as being that same behavior. If we define order as the absence of international conflict, for example,[9] then the argument that order decreases international conflict is as true as it is unenlightening.

My second reason for preferring an ideational rather than a behavioral definition of order is that the latter ties the existence of order to an ideal type of the behavior that it is said to produce. In international law this form of argument has come to be called the "perfect compliance fallacy" (Spiro, 2014). You've probably heard some version of this in casual conversation: "International law has no effect. Just look at all the countries that violate it!" (Most people who make this argument break the speed limit on a regular basis, but few if any of them would conclude that speeding laws are ineffectual or that there is no such thing as domestic order.) Continuing the previous example, if order were defined as the absence of conflict, then incidents of conflict would by definition represent a breakdown in order. As Professor Howard (2000, 2) points out, however, war is often instrumental in the maintenance of international order. Defining order as the absence of conflict thereby runs the risk of tautology or descriptive inaccuracy or—as difficult as it is to wrap one's head around it—both.

What Kissinger's definition omits, I think, is the crucial element of *anticipation* in international order. My colleague, Professor Randy Schweller (2014, 11), follows Bull's definition of international order but also offers a very helpful illustration of the concept:

> We call things orderly when the observer can grasp their overall structures and the meanings and purposes of those structures. Books arranged on a shelf according to author or subject—as opposed to size or color or merely heaped in a random pile on the floor—serve the functional purpose of selection. This is what Augustine meant when

he defined order as "a good disposition of discrepant parts, each in its fittest place."

The obvious part of the example is the fact that the arrangement of the parts is based on some purpose, as both Bull and Kissinger argue. The less obvious part is that the importance of that arrangement for understanding behavior lies in the fact that the *observer* can understand it *and act on it in anticipation of a particular outcome.* Someone looking for a book whose author's last name begins with an A can reasonably anticipate finding it toward the beginning of the bookcase rather than at the end. It's that anticipation that guides behavior, strictly speaking, not the existence of the consequence itself.

That anticipation, and the political arrangements that produce it, make a very compelling case that international orders can best be understood as *regimes.* Regimes are "principles, norms, rules, and decision-making procedures around which actor expectations converge in a given issue-area" (Krasner, 1982, 185). Regimes give structure to our expectations. We have a strong sense that we don't have to worry about Germany attacking France, for example, whereas in previous periods people were far less confident in that conclusion.

An analogy to everyday life helps to clarify what regimes are and how they arise. Those of us who use social media are aware of certain rules of etiquette that structure online interactions. Job seekers, or anyone likely to seek a job soon, should be aware that prospective employers will likely browse their public social media streams and should post accordingly. By now, most people have learned not to feed trolls—that is, not to respond to people who are making intentionally provocative statements in order to start arguments. And cases like that of Justine Sacco, whose offhand tweet prior to boarding a flight to South Africa inspired such instant and widespread outrage that she'd been fired from her PR job before she landed, have impressed upon social media users the speed with which offensive statements can produce a backlash with real-world consequences. Those consequences, in turn, have prompted campaigns against online shaming.

These rules or norms all arose spontaneously from the interactions of individual people in a context that's nearly devoid of formal oversight. Rules can also, of course, be imposed by formal institutions that govern interactions. Traffic laws help us coordinate our behavior and avoid running into one another. Property laws structure our expectations about whether other people will come into our houses and rummage through our stuff. Laws about assault convey the expectation that disagreements should be resolved without resort to violence. And so on.

More specifically, international orders are a subset of a larger category of regimes known as *security regimes*—"principles, rules, and norms that permit nations to be restrained in their behavior in the belief that others will reciprocate" (Jervis, 1982, 357). Examples of security regimes include not just international orders but also conventional alliances or treaties, which are far more common. They do not, to my mind, include purely anarchic systems, in which states are completely unrestrained in their behavior, although there's some disagreement on this point.[10] But they do include classical balance-of-power systems, such as the one that prevailed in Europe following the Peace of Utrecht in the early 1700s, which do include some constraints.[11] The motivation for security regimes is typically to improve member states' welfare, at the cost of some autonomy, by moderating the impact of anarchy on state behavior—in particular, by lowering the probability and efficacy of coercive behavior.[12] Again, the means utilized to achieve that goal are not always peaceful: defensive alliances, for example, seek to reduce the likelihood of the use of force via the threat of the use of force.

Three characteristics differentiate international orders from other security regimes. First, they are typically made up of more than a few countries. Second, the expectations that make up an international order are generally maintained by one or more of the most powerful countries in the international system. Finally, as Kissinger emphasizes, an international order is in part legitimated by a set of principles that could theoretically be applicable to the entire world.[13]

The latter point deserves a bit of explanation. By "legitimacy," I mean the capacity of a major power or powers to engender and maintain respect for, or even love of, the norms and principles that constitute an international order.[14] That respect matters, because people are more inclined to adhere to rules, norms, and principles if they believe that those rules, norms, and principles are appropriate and just. The same is true of domestic politics, of course: People are more likely to submit to governance if they buy into the legitimacy of domestic political institutions. For that reason, leaders of countries try to rely as much as possible on the legitimacy of the political system that they represent in order to govern—it's cheaper than coercion and a lot more effective. Thanks to leaders' focused efforts to reinforce legitimacy via national symbols, anthems, and so on, the governed often buy into domestic political rules, norms, and principles at a very fundamental level. To the extent that they do so, however, they limit leaders' freedom of action at the international level. It can be challenging for leaders to claim that one set of rules, norms, and principles legitimizes domestic governance but a *different* set is appropriate for international governance. For that reason, international orders tend to reflect the political ideologies of the major power or powers at their core, or at least not to be too obviously inconsistent with those ideologies, and those ideologies tend to be seen as broadly applicable by their adherents.

All of these considerations can be captured in a succinct definition:

> *International orders* are multilateral security regimes that involve one or more major powers and are legitimated by a set of principles that are potentially universal in scope.

Building a definition on layer after layer of abstraction in the way that I've just done can obscure as much as it reveals, so to make things a bit clearer, let me offer a quick example. The Concert of Europe, which I'll discuss in more depth in the next chapter, was an international order that existed in Europe following the Napoleonic Wars. Its essence as a security regime was the set of expectations that it created among major and minor powers alike regarding how the major powers would behave if revolution threatened the peace. The Concert was legitimated, superficially, via an appeal to the divine right of monarchs, though it is perhaps more accurate to say that it was legitimated by a common respect for the rule of law, which could most easily be imposed and maintained by monarchs (Schroeder, 1994, 529–530). It is much easier to say what the Great Powers stood *against*—liberal and nationalist revolution of the sort that sparked the Napoleonic Wars—than what they stood for. Establishing the common idea that revolution represented a grave threat to the peace that warranted a collective response clarified the major powers' likely reactions to popular uprisings and assured each that the others would not act without prior consultation.[15]

Given how much we tend to focus on power when discussing Great Power politics, it would be easy to overlook the role of legitimacy in producing and sustaining international order. To do so would be a mistake. Just as domestic political orders are founded on principles of legitimacy (popular sovereignty, the divine right of kings, the tenets of a given religion) and their policies must square with the norms and values associated with those principles, international orders are built upon a normative foundation that justifies the concession of political authority by the weak to the strong.[16] The ideals that make an international order legitimate in the eyes of constituent states need not drive policy, but policy must at least be minimally consistent with them: America justified propping up profoundly anti-democratic Third World dictators during the Cold War with the argument that the alternative, Communism, was worse still. An illegitimate order must either be ruled by coercion rather than consent, in which case it becomes an empire, or resign itself to internal strife and likely dissolution.

It is also important to note that international orders are often based on implicit or explicit theories of conflict. After all, it's impossible to come up with a plan for political order without a shared understanding of what the main threats to order are and how they can best be countered. At first, that diagnosis was fairly crude: The major threats to the European peace were believed to be attempts on the part of individual states to amass enough power to dominate the Continent. Accordingly, the first international order conceived of by the statesmen of Europe was rudimentary. It involved nothing more than the maintenance of a balance of power among the major states, such that none could reasonably contemplate the elimination of the others. This principle became formally embedded in European diplomacy in the Treaties of Utrecht, which put an end to the bloody and wide-ranging War of the Spanish Succession (1700–1714). The principle of the balance of power was invoked in that context to prevent the union of the French and Spanish thrones, but it was understood to be a general principle for preventing large-scale war. Small-scale wars were still understood to be natural and even in some instances necessary to preserve the balance. In fact, as the citizens of Poland discovered when their country was divided up and absorbed in the 1790s, war was often the means by which order was maintained. As Edward Vose Gulick (1955, 35), author of the classic work on the subject, put it, "Peace was no more essential to equilibrist theory than the barnacle to the boat."

Subsequent international orders evolved to counter more nuanced and varied threats. Historian Michael Howard, in a short but brilliant book called *The Invention of Peace,* describes five different schools of thought regarding international order in the nineteenth and twentieth centuries:

- The *conservative* school, which saw the French Revolution as an example of the threat that liberalism posed to the peace. Conservatives therefore sought to prevent future revolutions from having similar effects.
- The *liberal* school, which saw established European monarchies as the source of war and believed that representative government would bring mankind's natural virtue to the fore and ensure peaceful relations.
- The *nationalist* school, which saw the unjust incorporation of nations within the confines of empires and their division across national boundaries as a major source of war and saw the path to a lasting peace in widespread national independence.
- The *national socialist* school, which saw the world as a struggle for dominance among different peoples and sought to strengthen and protect the German *Volk* and either subjugate or eliminate the other peoples of the world.
- The *Communist* school, which saw modern industrial capitalism as the source of class conflict and war and viewed itself as capitalism's successor and mortal enemy.[17]

While the five schools of thought differed dramatically in their prescriptions for abolishing war, all but the Nazis believed in the eventual promise of a peaceful international order short of a world state. At the same time, none of them have had any compunctions about using war in order to achieve and, if necessary, maintain that order.

It is also worth emphasizing the fact that the shared willingness to cooperate in producing international order is often greatest in the immediate aftermath of its failure. As Professor G. John Ikenberry points out in his book *After Victory*, the establishment of an international order tends to follow the conclusion of a major war, when the need for it is most apparent and when missing or tentative borders make a re-drawing of the map possible. While Ikenberry (2001, 50) sees the rules and institutions that underpin order as serving "the long-term interests of the leading state," Professor Bruce Cronin (2010) argues that the war aims that help to create a wartime coalition, and that may or may not reflect the interests of the leading state, are more likely to inform postwar international order.

Finally, it is worth addressing one criticism that might have come to mind by now: the possibility that the relationship between international order and peace is tautological. It's possible that what we think of as international order is really just a situation in which there is no or little conflict among states, and the existence of conflict could be taken as evidence that order no longer exists. Fortunately, the concept of international order as it is typically used in the literature is conceptually distinct from conflict: as Professor David Lake (2009, 8) puts it, "Authority relationships are not devoid of coercion but are defined by the status or legitimacy of force when it is used." International orders very often legitimate the use of violence to preserve or extend order: Just as the Concert of Europe legitimated the use of force to put down revolution, the Western liberal order's responsibility to protect (R2P) doctrine legitimates the use of force to prevent such events as genocide and crimes against humanity.

The Sources of Legitimacy

While most students of the subject concur that enlightened self-interest is sufficient for the establishment of some degree of order, they differ with regard to how deep that order can become in the absence of the legitimacy that is conferred by a degree of normative or cultural affinity. Immanuel Kant is the philosopher most widely associated with the idea that universal peace can be achieved if all countries adopt what we now call a democratic form of government, regardless of culture.[18] In the modern era, the most compelling case for the argument that order can form among dissimilar states is the Concert of

Europe. Edmund Burke and, later, Karl Deutsch and his coauthors were more skeptical: Burke emphasized the importance of common language and tradition in generating a functional international order, while Deutsch et al. emphasized the importance of a collective "we-feeling" in producing what they refer to as *security communities*—collections of countries within which the use of large-scale violence in order to resolve differences is rare or unthinkable.[19] Similarly, Professor Daniel H. Nexon (2009) argues that the emergence of cross-cutting religious networks and the associated breakdown of shared values systematically undermined domestic order in early modern Europe, while Professor Charles A. Kupchan (2010, 2014) explores the importance of shared norms in the creation and maintenance of international order.

My own take is that for international order to function, legitimacy has to come from somewhere, but it can come from a great variety of places. My favorite example of an unusual source of legitimacy is that of the Easter Islanders, who for a time chose as their leader the first warrior to return from the mountains with the egg of a sooty tern.[20] The post-Utrecht balance-of-power system is evidence that legitimate order can be based on little more than power and self-interest, so I'm skeptical of the Burkean claim that cultural affinity is actually necessary for order. It's even easier to see the weakness of this argument if we apply it to domestic political order: Burke would have had an awfully hard time explaining the continued existence of Indonesia, for starters. That said, it seems more than likely that international order can be significantly enhanced when it is underpinned by shared values, norms, and principles of domestic political legitimacy.

The Role of Human Nature

But wait, you might be thinking. *Isn't international order just an outgrowth of the ongoing betterment of humanity and the ever-expanding circle of empathy that results from it? And if so (gotcha!), isn't this really just validation of the decline-of-war theorists' main argument?* Excellent question!

It's certainly true that international orders that are based on cooperation rather than coercion are, with a few exceptions, a recent invention by world-historical standards. To the extent that we can talk about human progress leading to more peace, in fact, the creation and promulgation of international orders that are not based on war is a very real achievement in human history.

That said, the relationship between international order and liberal Renaissance humanism is tenuous at best. The conservative school, most clearly associated with the post-Napoleonic Treaty of Vienna, was explicitly *illiberal* in nature—indeed, it saw Western liberalism as the problem rather than the solution. Communism saw itself, of course, as being the next stage in history after capitalism and believed Western imperialism to be a major threat to the peace. And no one in their right mind would call the National Socialists humanitarian or empathic. That doesn't mean that they can't all trace their origins to the Enlightenment, of course. But it does demonstrate that mainstream Western liberal thought is hardly the only tradition to have emerged from that period and far from the only one capable of producing international order.

As these examples demonstrate, the establishment of order does not depend on the progress of humanity or the spread of pacifistic norms of behavior. No lesser authority than Immanuel Kant makes exactly this point, and quite clearly, about domestic order. In his *First Supplement of the Guarantee for Perpetual Peace* Kant writes,

> The problem of organizing a state, however hard it may seem, can be solved even for a race of devils, if only they are intelligent. ... [I]t does not require that we know how to attain the moral improvement of men but only that we should know the mechanism of nature in order to use it on men, organizing the conflict of the hostile intentions present in a people in such a way that they must compel themselves to submit to coercive laws.

By the same token, international orders can occur, not when people become pacifists, but when Great Powers choose to forgo some of their interests because they realize that it is advantageous to do so. The Great Powers may choose to create and sustain international order because they realize that long-term cooperation is more valuable than the short-term, unilateral gains that would preclude it, and smaller powers may sign on because they agree. Altruism, humanitarianism, and nonviolence are not necessary for cooperation to occur in an anarchic world: Great Powers can produce the international order that constrains their actions solely because it is in their interest to do so.

Moreover, even in cases in which international orders are imbued with the norms and principles of Western liberalism, their implications for peace are somewhat more convoluted than their proponents would have us believe. I now turn to a more in-depth discussion of this point.

Order and the Sources of Systemic Conflict

Now we get to the heart of the matter: the relationship between international order and international conflict. Above, I've argued that international orders promote internal peace and external conflict, especially in the presence of competing international orders. How exactly do they do so?

The answer, in brief, is this: International orders have an impact on the issues that drive conflict and on the calculations that states make when deciding whether to fight over a given issue. Even if states are rational and war is more costly than a negotiated settlement, a state's calculus can still favor war in the presence of *information problems* or *commitment problems.* International orders can have an impact on the *issues* that lead to war and on states' *calculations* about war with the aid of a dense network of interconnected political institutions, as most notably the Western liberal order has done; or they may do so in the virtual absence of international institutions, as Professor Mitzen (2013*b*) and others have argued was true of the Concert of Europe. Among the states that comprise international orders, the changes produced by international order generally promote peace; outside of (and especially across) international orders, those changes promote conflict.

This answer paints a very rationalist picture of the origins of war. To anticipate a likely objection, I'm not arguing that all wars begin for rational reasons—far from it. I do believe that people weigh the costs and benefits of going to war, but how much those costs and benefits will matter in the presence of ancillary considerations, psychological biases, risk attitudes, situational effects, and so on is hard to say. I focus on the rational part of the decision to go to war, not because I think it dominates the others, but because it is the part that international orders are typically designed to influence.

With that in mind, let me unpack that answer a bit more.

Issues and War

The most straightforward part of war is that it is generally *about* something—that is, wars are generally fought to resolve some issue or another. The issues that prompt most wars fit fairly well into one of a fairly manageable number of categories. Quite a few are about territory, for example, or the creation or dissolution of countries, or the defense of those countries' integrity, or dynastic succession, or the defense of co-religionists or co-nationals. Granted, the history of warfare shows that countries can fight over some pretty weird things: Some armed clashes between the British and the Dutch in the 1600s, for example, were fought over access to nutmeg.[21] However idiosyncratic the *casus belli* may seem, however, there generally is one.

The attempt to understand warfare as a function of the issues that prompt it has, accordingly, become a fairly common impulse among scholars.[22] The Australian historian Geoffrey Blainey's book *The Causes of War* is one of the best-known works in this tradition. Blainey covers very general causes, such as miscalculation, as well as very specific ones (his evocatively named "Death Watch and Scapegoat Wars," for example). He also underscores the importance of more mundane factors, like weather, that most theorists of international conflict have overlooked until very recently. Similarly, Professors Robert I. Rotberg and Theodore K. Rabb explore the international, domestic, and individual-level roots of major conflict in a unique blend of theoretical and historical essays entitled *The Origins and Prevention of Major Wars.*

In his book *Peace and War,* Professor Kalevi Holsti took on the ambitious task of cataloguing and categorizing wars since 1648 by cause, with an eye toward understanding trends in the issues that provoke armed conflict. His findings (Holsti, 1991, 307, Table 12.1) are illustrated in Figure 7.1. As we can see from the figure, territorial clashes have been a persistent cause of war since the Treaty of Westphalia, while some causes of war (commerce and navigation, succession) have dwindled and others (national liberation, the maintenance of the integrity of the country or empire, the composition of government) have become more prominent.

Most of the scholars who focus on the issues that prompt international conflict, however, focus on depth rather than breadth. The

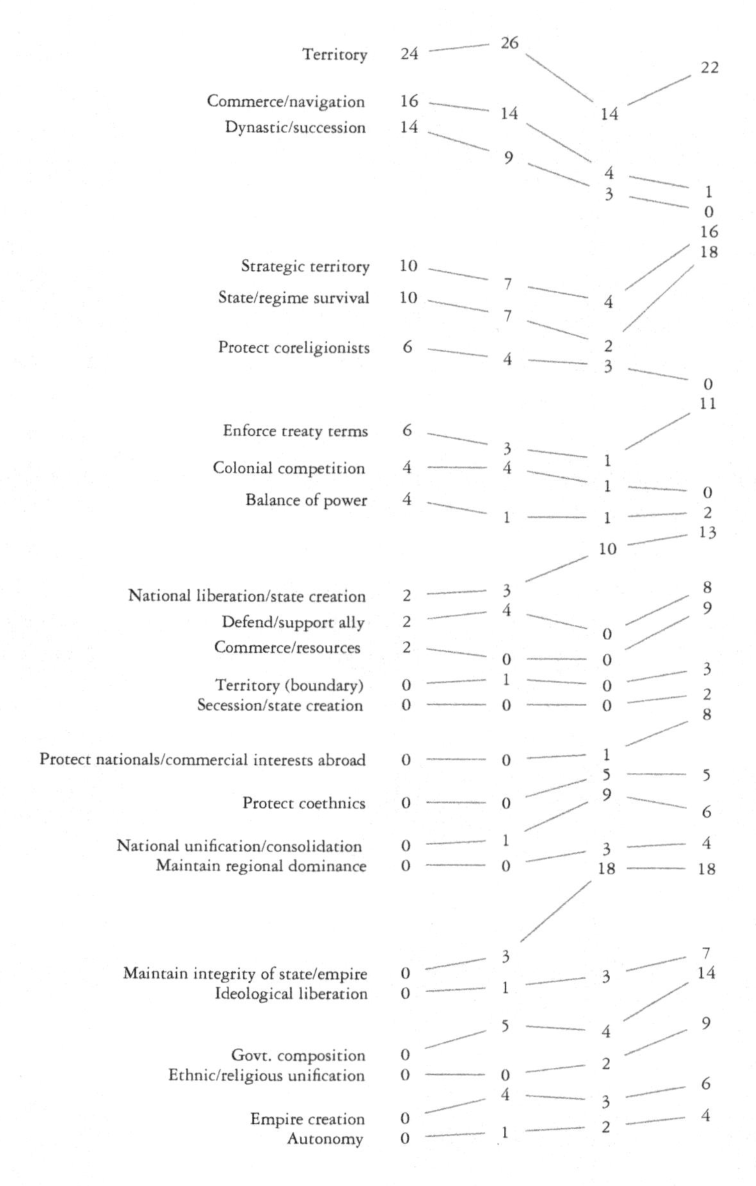

FIGURE 7.1 Issues at the heart of armed conflict, 1648–1989. Source: Holsti (1991).

majority explore the logic of and data pertaining to a single conflict-causing issue. Many do so for the better part of their careers. As their numbers have grown over time, scholarly work that either theorizes, refines theories about, or tests theories about the origins of conflict has come to make up a large percentage of research in the conflict studies subfield of political science. Those causes of conflict include issues such as the struggle for power,[23] arms races and conflict spirals,[24] ethnicity and nationalism,[25] domestic political regime type[26] and leadership change,[27] economic interdependence and trade,[28] territory,[29] climate change-induced scarcity,[30] and so on.

As the above discussion suggests, some issues are more prominent causes of war in one era than in another. Often, I suspect, those changes in prominence are a function both of the times and of small numbers, as I noted in chapter 6. Given the rarity of disputes in general, it shouldn't be a surprise to see a lot of fluctuation over time.

In many ways, Figure 7.1 supports this intuition. Dynastic wars, or wars of succession, have decreased dramatically as a cause of war since the seventeenth century, largely because succession no longer serves either to cement territorial holdings legitimized by continuous bloodlines or to create de facto alliances or long-standing allegiances among the Great Powers. Wars of national liberation and wars to maintain the integrity of the state or empire both skyrocketed to prominence in the wake of the French Revolution and the Napoleonic Wars, which brought national independence into conflict with divine right when it came to determining the boundaries and existence of countries. Similarly, the protection of coethnics and ethnic or religious unification went from being literal nonissues prior to the French Revolution to comprising almost 15% of all wars following World War I.

The Impact of Order

How might the presence of international order have an effect on the prevalence of issues that lead to war? First of all, international orders can mitigate or eliminate the issues that have caused conflict in the past. In fact, those that are constructed in the wake of large-scale wars are typically explicitly designed to do so. This goal can most explicitly be seen in Woodrow Wilson's Fourteen Points, a statement of war aims designed to ensure a lasting peace after World War I. It's true that, as Professor Holsti points out in his study, the diplomats and politicians who craft new international orders generally have a far better grasp of the issues that have produced war in the recent past than they do of the issues that will generate war in the future. But that generalization may be true at least in part because those diplomats and politicians succeed in addressing the issues that had previously led to war.

At the same time, agreement on the principles of legitimacy that underpin international order can take other war-causing issues off the table. To take an extreme example, an international order based on political democracy more or less eliminates the incentive for wars of royal succession. More subtly, agreement on the principles and values that underpin order can reduce or eliminate conflict based on violations of those principles and values. To take an example from the realm of domestic political order, armed conflict broke out in the aftermath of the 2010 Presidential election in the Ivory Coast after the incumbent, Laurent Gbagbo, refused to accept his loss in the election. United Nations forces were tasked with protecting the newly elected President, former Prime Minister Alassane Ouattara, and preventing both escalation and human rights abuses. When Gbagbo's forces unwisely fired on the peacekeepers' base with heavy weaponry, UN and French forces retaliated unambiguously, and within two months Outtara had been officially installed as President.[31] While it's not impossible to imagine a situation like this breaking out in a country with more established democratic norms of political contestation—Ivorian opposition parties were only legalized in 1990—such a clash seems far less likely in, say, modern-day Germany or France.

Moreover, the existence and development of an international order is often characterized by a positive feedback process in which the normative underpinnings of political legitimacy are deepened and reinforced among member countries.[32] To the extent that this process occurs, the persistence of international order should make conflict over issues related to those norms less and less likely over time.[33]

On the other hand, international order can actually produce new issues over which states inside the order will fight with states outside of it. International order depends on legitimacy, and once an international order has been created that legitimacy must be maintained.

Unfortunately, one of the most convenient ways to enhance that legitimacy is via external conflict. In American foreign policy, following Professor Mueller's (1970) landmark article, this phenomenon is known as the "rally 'round the flag" effect. While international orders don't typically have flags, they nevertheless exhibit rally effects, as when the Soviet Union responded to the American announcement of the Marshall Plan in 1947 by calling a meeting of international Communist parties at Szklarska Poręba, Poland, to establish an "anti-imperialist democratic camp" and legitimate the Soviet Union's central role in the international struggle against capitalism.[34]

Because the orders that they create are based on principles of political legitimacy and those principles are often mutually exclusive, competing international orders inherently represent a new issue—political legitimacy—over which countries can fight, either directly or in peripheral areas. The most obvious recent case that matches this description is the Cold War: The Soviet Communist world order was organized around the notion that Western capitalism was a fundamentally immoral organizing principle for society and had to be superseded by socialism, and the liberal West, not unreasonably, saw Communism as a threat to its basic principles. Because their legitimating political ideologies were incompatible and each side feared the expansion of the other, the question of political ideology came to the forefront in conflicts around the globe.

Maintaining security commitments within orders can also become a conflict-producing issue that would not have existed in the absence of order. Typically, the Great Powers at the heart of an order have a greater obligation to provide security when called upon to do so than they would otherwise have had, which can lead to the expansion of conflict in crises with external actors.[35]

The existence of competing international orders can also prompt countries or rebel groups to couch their policy goals in terms of the principles of legitimacy that underpin one of those orders in an attempt to garner Great Power support, and for the political leaders of the Great Powers to accept and amplify that framing in order to justify intervention. This was clearly the case in Nicaragua in the 1980s, where the right-wing Contra rebels made themselves out to be, in President Reagan's famous words, "the moral equal of our Founding

Fathers and the brave men and women of the French Resistance." Judging by the Contras' human rights record,[36] this statement was either false or a scathing indictment of our Founding Fathers and the brave men and women of the French Resistance. But posing as committed democrats was politically very useful to the Contras in attracting American support.

The Calculus of War

The existence of issues worth fighting over, while crucial to our understanding of the origins of war, is not a complete explanation of war. As Professor James Fearon (1995) pointed out in his widely cited essay "Rationalist Explanations for War," the great majority of the issues that countries fight about could have been settled by negotiation. Settling them by negotiation would have been unambiguously less costly than war, even for the victor: If everyone understood who would win a war, who would lose it, and how badly it would be lost, everyone could simply jump ahead to the postwar settlement without actually having to fight over it.

Professor Harrison Wagner pushes this logic further by arguing that, when no agreement short of war is possible, warfare is the means by which such an agreement is reached. As Wagner (2000, 469) succinctly puts it, "If fighting is expected to lead to an agreement, then fighting must be considered part of the bargaining process and not an alternative to it." In this understanding of warfare, war is a form of costly negotiation. War begins due to a lack of information, and the point of fighting is to reveal the information needed to arrive at a settlement. A state with an incentive to bluff or pretend to be stronger than it is, for example, will soon have its weakness revealed on the battlefield and be forced to accept a settlement that reflects its weakness.

The fundamental insight underlying this perspective is that, far from being an alternative to bargaining, warfare is an extension of it—a form of armed negotiation that allows countries to resolve the uncertainty that made a negotiated settlement possible. While this understanding of war had been suggested before,[37] the connection between Fearon's insights about pre-war negotiation and Wagner's perspective on bargaining during war suggested a new theoretical understanding of warfare. Rather than a process of negotiation that led to a costly lottery—essentially, a crapshoot—the bargaining model suggested, as Clausewitz emphasized, that "war is nothing but the continuation of policy with other means."[38]

The bargaining model of war probably seems pretty bizarre to people outside of the discipline of political science. (Well, that's not quite true: economists would find it perfectly normal.) To citizens of countries that start wars, it certainly seems as though there's a lot of passion involved, which makes this cold, calculating perspective seem sort of odd. To people on the receiving end of an attack, the enemy can seem far too irrational to step back and weigh the merits of a negotiated settlement. We're humans, after all, not Vulcans—do we really carry out these sorts of calculations when it comes to starting a war?

It's actually not too much of a stretch to conclude that we do something of the sort. For one thing, countries do negotiate all the time—even the ones that seem less rational from one's own point of view. Those negotiations often have to do with issues that matter a lot to the countries that are involved. We often don't notice them precisely *because* they succeed: Countries often do manage to negotiate quiet, effective settlements in situations that could have turned violent but didn't.

Take, for example, the unbelievably bloody and violent civil war that did not mark the division of Czechoslovakia into two separate countries in 1993. The ethnic hatred that did not drive Czechs to massacre Slovaks never made international news, just as the tragic damage that was not inflicted on the ancient Slovak city of Bratislava never prompted international relief efforts. The non-refugee crisis that ensued never put a strain on neighboring Austria and Hungary... well, you get the point. What in fact *did* happen was that the Slovak National Council passed a Declaration of Independence in July 1992, the Czechs and the Slovaks promptly negotiated a dissolution, and in January 1993 Czechoslovakia quietly disappeared into the history books.

In any event, the point of Fearon's essay is not to argue that we are all emotionless automata who can dispassionately game out the likely consequences of war and arrive at a tidy settlement without having to pay war's costs. His point is that leaders who are contemplating war always have the option of seeking a negotiated settlement instead, and given that war is a very dangerous and destructive enterprise they have to give that option serious consideration. Do they all have to be perfect at predicting the likely outcome? No, not at all. But when we realize that the last step between a disputed issue and war is the decision to choose war over negotiation, it changes our fundamental orientation about the causes of war. The existence of issues worth fighting over is necessary for war, but it's not sufficient: Someone must also choose either to abandon negotiations or to forgo them entirely.

That altered perspective, in turn, leads us to wonder why some countries fail to come to a negotiated settlement at this crucial juncture. What separates the Czechoslovakias from the Yugoslavias? Why is it that some serious disagreements over major issues are resolved peacefully while others devolve into war?

Information Problems

Fearon highlights two main kinds of problems that can prevent countries from reaching a negotiated settlement. The first are *information problems*: War can occur if the two sides have different information (about, say, the capabilities of their militaries or the resolve of their people) but cannot convey that information to one another credibly. If they had different information but *could* convey it to one another, they'd end up with roughly the same estimate of who would win a war and by how much, and they'd be more likely to reach a negotiated settlement rather than fighting. But sometimes the leaders of countries find themselves in situations in which they simply can't believe one another or have reasons for not revealing information to one another. Because the specifics of a country's military capabilities are often among its most closely guarded secrets, for example, it's not difficult to imagine situations in which leaders could develop irreconcilably different ideas about their country's chances of winning a war.

The Falkland Islands war of 1982 is a good example of how information problems can produce war. The Falklands, a cluster of small islands that lies about 300 miles east of Argentina, have been a British overseas territory since 1833 but had previously been administered by the United Provinces of the Rio de la Plata, of which Argentina was a part. After achieving independence, Argentina maintained its claim to the Falklands, and it began to press those claims with increasing seriousness in the 1960s and 1970s. In 1982, the military junta in charge of the country, encouraged by the gradual erosion of Great Britain's ability to project military power, Argentina's growing friendship with the United States, and the Argentine military's recent acquisition of advanced military hardware (in particular, the Exocet missile), concluded that the costs of going to war over the Falklands would be prohibitively high for the British. Accordingly, they launched a surprise attack to re-take the islands in the hopes that possessing them would give Argentina a decisive advantage in the negotiations that were sure to follow. There was a risk that the British would respond with force, of course, but the junta decided that the risk was low enough to justify the gamble.

Unfortunately for them, the gamble didn't pay off. The British, under Margaret Thatcher, proved more than willing to prosecute a war against Argentina, even requisitioning civilian vessels like the *Queen Elizabeth II* to transport soldiers to the South Atlantic. The United States, which the junta had hoped would remain neutral, provided material support to the British and banned weapon sales to Argentina. After just over two months of fighting and nearly a thousand casualties, British forces retook the islands.[39] The fact that the junta chose to attack without warning limited the ability of the British and the Americans to warn them that they were about to make a very bad decision. When it became clear that an invasion would take place within twenty-four hours, President Reagan spent fifty minutes on the phone with President Galtieri but failed to convince him of British and American resolve (Gwertzman, 1982). In fairness, it wasn't unreasonable to suspect that they were bluffing. For that reason, Argentine uncertainty about the value of the Falklands to the British and the value of British friendship to the Americans was pretty much unavoidable; that uncertainty made it reasonable for Argentina to gamble that the British would not mount a costly military response to an invasion and wouldn't receive significant American assistance if they did; and that turned out to be a bad gamble.

The Impact of Order

There are a variety of ways in which international orders and the institutions that underpin them change states' information environments in a manner that is conducive to peace. First, and perhaps most obvious, one of the most common means by which international orders ameliorate information problems is via the creation of international institutions that enhance the quality of information available to the countries in the international system.[40] The 2015 Joint Comprehensive Plan of Action between the Islamic Republic of Iran and a group made up of the P5+1 countries[41] and the European Union is an excellent example of an institutional attempt to solve an information problem by reducing Iran's stockpiles of nuclear material, limiting its production of nuclear material, and implementing credible verification measures. The long-term effect of the increased transparency of Iranian nuclear production, it was hoped, would be a general reduction of tensions with the West and with other countries of the world.

International orders also promote peace because the norms and principles upon which they are based lead to rules of international conduct that the order promotes and that the leading states enforce. The resulting reduction in uncertainty has a pacifying effect among member states. A good example of this sort of dynamic occurred in the Suez Crisis in 1956, when Egyptian President Gamal Abdel Nasser nationalized the Suez Canal and the United Kingdom, France, and Israel conspired to invade the Sinai Peninsula in the hopes of regaining control of the canal and ousting Nasser. The United States, which at the time hoped to create a NATO-like security structure in the Middle East to forestall Soviet advances, used overwhelming financial pressure to put a stop to the invasion and force its own NATO allies and Israel to withdraw. The relevant rule of international conduct, in this case, involved not disrupting the development of a Middle Eastern security organization that would have hindered Soviet interference in the region, and when its allies violated that rule by attacking the country that was meant to be the centerpiece of that organization, the United States took remedial action—even going so far as to vote against Britain and France, and with the Soviet Union, at the United Nations.[42]

There is a substantial body of scholarly evidence to support the related claim that information provision by international institutions enhances the prospects for peace. Recent scholarship has shown that joint membership in information-providing international organizations reduces the probability of conflict onset.[43] Research has also demonstrated that peace agreements that require monitoring, verification, and the provision of military information reduce the chances that war will recur (Mattes and Savun, 2010). Finally, the security coordination that often results in international orders is itself a source of information and threat reduction (Lake, 2009, 101). Joint military planning, combined exercises, and integrated command and control structures all have the effect of building confidence in one another's intentions and significantly decreasing uncertainty about one another's capabilities.

At the same time, international orders can create information problems where none existed before, and those information problems can lead to conflict with states that are not part of the order. For instance, a security commitment that arises from order can produce a challenging dilemma. If that commitment is too weak, it serves little purpose. If it is too strong, however, it may encourage smaller states to take risks because they know that the more powerful state or states at the heart of the order will bear the costs—a situation known to economists and insurance companies as "moral hazard." In order to avoid moral hazard, the central state or states must often create a degree of uncertainty in their commitments. Unfortunately, that uncertainty makes potential opponents uncertain as well and generates information problems that can lead to war.[44]

The case of China and North Korea is an excellent example of this dynamic. The 1961 Treaty of Friendship, Cooperation and Mutual Assistance between the two requires China to "render military and other assistance by all means at its disposal" if North Korea is attacked. The treaty serves as a useful deterrent, but it also emboldens North Korea to take risks that it might otherwise not take. China maintains a degree of ambiguity by retaining the right to determine what constitutes an attack on North Korea. That fact, as well as China's uncertain leverage over its protégé, creates substantial uncertainty in the minds of South Korean and Western politicians when they try to engage North Korea.[45]

To make things worse, the uncertainty created by the need to avoid moral hazard is even greater in an international order than it is in an ordinary alliance, simply because more states are involved. This fact can generate the pathologies that Professors Thomas Christensen and Jack Snyder (1990) memorably referred to as "chainganging" and "buckpassing." In the former situation, more and more countries are drawn into the conflict by virtue of their ties to the original combatants. In the latter, each attempts to "pass the buck" of intervention to the others, and no one intervenes. Because an international order is typically made up of many states, and those states can all either chaingang or buckpass if one of them becomes involved in a conflict, it becomes very difficult to anticipate the outcome of such a conflict with any degree of certainty. World War I, in which alliances and treaties ensured that all of the major European states would be dragged into a conflict between Austria-Hungary and Serbia, is the canonical example of chainganging; the period prior to World War II, in which the major states were reluctant to stand up to Nazi Germany, is the typical example of buckpassing.

Commitment Problems

Professor Fearon's second main explanation for war involves barriers to settlement known as *commitment problems.* "Commitment problems" is one of those rare technical terms whose meaning actually corresponds pretty well to its meaning in real life. If you've ever dated someone with commitment issues, you've seen firsthand the phenomenon of someone who likes to be with you *now* but can't really commit to wanting to be with you in the longer term. They might have a fear of intimacy that makes them wary of the person they would become if the relationship became more serious, or they might believe that settling down would make them resent their lost freedom. Whatever the reason, they cannot commit to a long-term relationship because, even though they are happy in the short term, they believe that they wouldn't be in the long term.

Countries don't fear intimacy, of course. But they do face situations in which they're fairly sure that a commitment made today won't be one that they or their partner will be willing to uphold in the not-too-distant future. In the most prominent article to date that explores the logic of commitment problems and war, Professor Robert Powell (2006) lists three common scenarios in which countries might encounter commitment problems:

- *Rapid shifts in the balance of power,* which create an incentive for the rising power (call it country A) to renegotiate or renege on an agreement in the near future. Knowing this, country B can reasonably conclude that no enforceable agreement is possible today. The Peloponnesian War, which Thucydides described as having its origins in "the growth of Athenian power and the fear which this caused in Sparta" (Thucydides, 1972 [431 B.C.E.], ch. 1, sec. 23), is probably the most famous example of this dynamic, though historian Donald Kagan (1969) argues convincingly that it's deeply problematic; German fears of being eclipsed by the growth of the Russian military prior to World War I is a better one (Herrmann, 1996; Clark, 2013, 326–334).

- *First-strike or offensive advantages,* which create a situation in which the short-term advantage to striking first outweighs any advantage that might be obtained by bargaining. The Japanese attack on the American Pacific Fleet at Pearl Harbor exemplifies this sort of commitment problem.
- *Bargaining over issues that represent a source of military power,* in which case the settlement itself, if successful, would change the balance of power enough to prompt one of the countries involved to renege. The Munich Agreement of 1938, in which the major European powers ceded the Czech Sudetenland to Nazi Germany, is an example of such a situation: Because the Sudetenland contained the bulk of Czechoslovakia's border defenses, the agreement rendered the country helpless against the Nazi invasion that came a few months later. (In this case, a negotiated settlement actually was reached, of course, and its universal condemnation as one of the stupidest agreements in the history of diplomacy underscores the folly of signing agreements that create commitment problems.)

It's important to note that commitment problems can arise even in the absence of uncertainty or private information. What that means, in turn, is that the dynamics of such wars are very different. If a true commitment problem exists, no negotiated settlement can resolve it—and no amount of information gained during the course of fighting is likely to change that fact. For that reason, it is difficult to imagine wars that stem from commitment problems being resolved quickly or with little bloodshed. Indeed, as Professor Alex Weisiger (Weisiger, 2013) shows, wars that stem from commitment problems are unusually long and deadly.

The Impact of Order

There is substantial evidence that the institutions that typically underpin international orders can decrease conflict by reducing the severity of commitment problems. Professor Megan Shannon, for example, demonstrates that international organizations with security mandates that have the ability to intervene diplomatically in members' disputes increase the probability that a dispute will be settled peacefully.[46] Similarly, Professors Michaela Mattes and Burcu Savun (2009) show that institutions contain fear-reducing and cost-increasing provisions to minimize commitment problems in postwar settlements, while Professors Sara McLaughlin Mitchell and Paul Hensel (2007) demonstrate that they allow leaders to save face with constituencies when accepting unpopular settlements.

International orders can also help to solve commitment problems by acting to reduce the severity of the worst-case scenarios that drive such problems. This can occur when the leading state agrees to limit its own behavior in exchange for a security commitment from subordinate states (Ikenberry, 2001, 34), or when multiple states commit in advance to come to the aid of a member state that is attacked. NATO's Article V, for example, which states that an attack on one member state is an attack on all, helps to reassure member states that their people will be protected and their sovereignty preserved. With such an assurance in hand, member states can be more sanguine than they would otherwise have been about concluding negotiated settlements to which they fear an adversary might not be able to commit.

Finally, Professors Anja Hartmann and Beatrice Heuser (2001, 247–248) argue that the shared norms and principles of political legitimacy that underpin an international order can provide not just information but a sense of community, or shared identity, among member states that reduces their fears about the use of force and moderates concerns about violations of the rules of war. While this outcome certainly provides those states with more information about one another, it also has the effect of lowering their estimates of the likely costs of future warfare, which helps to ameliorate commitment problems.

Commitment problems can also be exacerbated by the existence of international order, however. To the extent that order provides stability for member states, it can increase the probability that outside states will continue to interact with them in the future, thereby lengthening the "shadow of the future" that can improve the prospects for cooperation. As Professors James Fearon (1998) and Dustin Tingley (2011) note, however, a longer shadow of the future can actually make commitment problems worse because states have to worry about the enforceability of settlements over an even longer period.

Moreover, Professors Hartmann and Heuser (2001, 247–248), who argued above that shared identity within orders can foster peace, also argue that the establishment and solidification of international order deepens identity divisions with external states, producing precisely the opposite effect.

Table 7.1 summarizes the mechanisms I've just described that produce peace within orders and conflict among or outside of them. By compiling such a substantial list, I don't mean to imply that all of these dynamics will always occur in every case whenever an international order or orders exist. Some leaders of major states engage in "rally 'round the flag" behavior, while others don't, for example, and states can come to mutually agreeable settlements even in the face of significant uncertainty and commitment problems.

Rather, the takeaway should be this: When it comes to relations within an international order, the metaphorical deck is stacked in favor of peace—so much so that solitary international orders most likely eliminate more conflict than they produce. But when it comes to relations between an international order and peripheral states or (especially) between or among multiple international orders, the deck is often stacked in favor of conflict.

Conclusion

Although the details of the arguments that I've laid out above can get a bit convoluted, the upshot is straightforward: When it comes to peace,

TABLE 7.1 Summary of some mechanisms by which international orders decrease the probability of internal conflict and increase the probability of external conflict.

	Within Orders	Among or Beyond Orders
Issues	Orders that are based on general peace settlements institutionalize the resolution of issues that led to war.	Desire to enhance legitimacy tempts leaders to engage in external conflict.
	Agreement on fundamental principles of legitimacy upon which order is founded removes cause of conflict.	Clash of fundamental principles of legitimacy becomes new cause of war, especially with competing orders.
	Positive feedback reinforces and deepens legitimating principles over time, further decreasing probability of conflict.	Security commitments inherent in international orders increase odds that dominant powers will come to aid of subordinate ones in crises with external actors? Third parties couch conflicts in terms of legitimating principles in order to attract Great Power support.
Information Problems	Leading state has fewer security concerns when dealing with subordinate states. Information provided by institutions (via monitoring, alliances, international agreements, etc.) reduces uncertainty.	Twin fears of entrapment and abandonment/defection prompt Great Powers to create strategic uncertainty with client states and hence with their potential opponents.
	Orders comprise, and leading states enforce, rules of international conduct, which decreases uncertainty.	Aggregation of states into international orders produces chainganging and buckpassing dynamics within orders, increasing the amount of irreducible uncertainty.

(*continued*)

TABLE 7.1 Continued

	Within Orders	Among or Beyond Orders
Commitment Problems	Orders generate institutions, which seek to defuse conflict by mitigating commitment problems.	Longer time horizons produced by stable international orders exacerbate commitment problems
	International orders reduce the severity of worst-case scenarios that drive commitment problems.	Increased involvement of major powers in peripheral conflicts and in undermining settlements decreases ability of peripheral actors to make credible commitments.
	Shared norms and principles of legitimacy within orders foster shared identity, which reduces fears of the use of force and concerns about violations of norms of warfare.	Diverging norms and principles of legitimacy across orders foster distinct identities, which increase fears of the use of force and concerns about violations of norms of warfare.

international orders are a mixed blessing. There are a lot of reasons to believe that they will reduce conflict among member states. There are also a lot of reasons to believe that their relations with states outside of their own order may be more conflictual than they otherwise would have been.

To the extent that the decline-of-war theorists have addressed the issue of international order, their claims have been not so much wrong as incomplete. It's true that a strong normative commitment to peace drove state leaders to seek solutions to the problem of war—but those norms did not translate unproblematically into peace. It's true that Europe has been a remarkably peaceful place since World War II—but it spent most of that period divided between two international orders whose proxy wars ranged across the globe. It's true that the Western order is a liberal one—but other orders, which have also succeeded in suppressing international conflict, have not been. Expanding the scope of our view, even to the early nineteenth century, provides invaluable perspective on these issues.

In the last substantive chapter of the book I will briefly explore the historical record from the past two centuries to evaluate the general arguments that international orders reduce internal conflict but increase external conflict, especially with each other. Once I've done so, I return to the data on militarized interstate disputes to see whether patterns of international order are a more useful guide to conflict initiation than the straightforward decline-of-war hypothesis.

8

History and International Order

IN THE PREVIOUS CHAPTER, I offered an argument to explain the patterns of conflict that emerged from chapter 4. That argument hinged on the complicated nature of international order. I argued, first, that international orders quell conflict within their boundaries; second, that states in international orders can engage in more external conflict than they otherwise would and that multiple international orders are especially prone to clash; and third, that the war-suppressing effects of international order decrease with scale.

In this chapter I turn to the historical record to evaluate the plausibility of these hypotheses. I explore four major periods—nineteenth-century Europe, the interwar period, the Cold War, and the post–Cold War period—with an eye toward both patterns of conflict within and across international orders and the mechanisms (issues, information problems, and commitment problems) that I've argued have influenced them. A comprehensive diplomatic history of the past two hundred years would take up another whole book or two, but those books have already been written in part and in whole,[1] and in any event writing a new one is hardly necessary. Even a brief outline will suffice to warrant the claim that these patterns are far more consistent with international politics than they are with a growing trend toward humanitarian nonviolence.

At the end of the historical summaries, I turn back to the Militarized Interstate Dispute data as a check on my interpretation of these historical periods and as a more formal test of my arguments about the relationships between international order and international conflict. The results tell a fairly compelling story. The most peaceful groups of states on record are those that belong to international orders, though their relationships with third parties are more disputatious, and relationships across multiple orders can be extremely prone to conflict.

Nineteenth-Century Europe

The Concert of Europe

The four decades following the Napoleonic Wars were, by a significant margin, the most peaceful period on record in Europe. This is no coincidence. The map of Europe that was drawn by the Great Powers during the Congress of Vienna kept the center of the Continent weak and divided enough not to represent a threat, strong enough to resist a resurgent France, and reasonably independent of outside influence. The signatories wisely avoided a powerful Grand Duchy of Warsaw that would almost certainly have been a vehicle for extending Russian influence into central Europe, and they defused a serious disagreement between Prussia and Austria regarding the borders and composition of German territories in the same region. They also, after some initial hesitation, readmitted France into the ranks of the Great Powers. Austria and Prussia were bolstered with additional territory, and a German Confederation was established under Austrian leadership, in part to forestall future French expansionism. The overall result was that Russia, Prussia, Austria, France, and the United Kingdom were left in a position to counterbalance one another if necessary, but the fact that the weakest of them, Prussia, was located in central Europe along with German statelets like Hanover, Saxony, Bavaria, Württemberg, and Hessen-Darmstadt greatly reduced the other Great Powers' concerns over one another's immediate designs.

At the same time, the Great Powers benefited from an unusual degree of ideological solidarity. The French Revolution and the bloody wars that followed went a long way toward discrediting the liberal idea of constitutional governance. Austria and Prussia were anything but liberal, and when Napoleon was driven from power the conservative Bourbons returned to the throne of France. The United Kingdom, prior to the Reform Act of 1832, was only marginally democratic at best, and its foreign policy in the immediate aftermath of the French Revolution was actively anti-democratic: The British conspired with the Austrians to undermine the Sardinian constitution, worked to re-implement the three eighteenth-century partitions of Poland, and supported the Bourbon restoration in France. Only Russia, under the leadership of the mercurial Tsar Alexander I, flirted with constitutionalism in Poland and France before an uprising in Chuguev in 1819 and the mutiny of the Semenovski Guard in 1820 convinced him that liberalism posed a genuine threat. Once the dust settled after the end of the Napoleonic Wars, then, the Great Powers were unified in opposition to both liberal revolution and constitutional governance.

While some historians focus almost exclusively on the anti-revolutionary nature of the Concert[2] or the balance of power that underpinned it (Gulick, 1955; Slantchev, 2005), to do so is to miss its larger significance: The Concert was a focused and united attempt to construct a durable international order. It is true that the Vienna settlement created both a territorial status quo that minimized the Great Powers' fears of one another and codified a rare degree of consensus on the norms and principles that supported domestic political order—two conditions that are highly conducive to peace (Kissinger, 1994, 77). But the essence of the Concert was an established and acknowledged security regime in which the Great Powers agreed to forego the single-minded pursuit of their individual interests in order to cooperate in managing European security affairs, while the smaller powers conceded to them the right to do so (Schroeder, 1994, 578; Howard, 2000, 42–43; Kissinger, 2014, 66–67). There was nothing altruistic in the establishment of international order: as Sir Francis Harry Hinsley (1963, 197) wrote,

> [t]he impressive thing about the behaviour of the Powers in 1815 is that they were prepared, as they had never previously been prepared, to waive their individual interests in the pursuit of an international system. This fact is not rendered any less impressive by the recognition that they were prepared to waive their individual interests because it was in their individual interests to do so.

It is difficult to overstate the significance of the Concert in historical perspective. Whether an international order in any meaningful sense had existed prior to the Napoleonic Wars is debatable; to the extent that it did, the behavior that it prescribed was hard to distinguish from simple, self-interested aggression. The titles of two prominent works of history that cover the early 1800s, Professor Sir Michael Howard's *The Invention of Peace* and Professor Paul Schroeder's *The Transformation of European Politics,* give some sense of the magnitude of the change represented by the Vienna system.

Because the Concert of Europe was so thinly institutionalized—it had no fixed schedule, no legislative body, and no centralized meeting place—and because its origin lay in the desire to maintain the status quo against revolution and war, it is easy to mistake changes in that status quo for the degradation of the Concert itself.[3] So, for example, while the Great Powers met in Congresses at Aix-la-Chapelle (1818), Carlsbad (1819), Troppau (1820), Laibach (1821), and Verona (1822) to address residual threats to the peace in the aftermath of the Napoleonic Wars, the series of Congresses stopped abruptly at that point. When the July Revolution of 1830 ousted Charles X of France and reintroduced popular sovereignty as the foundation of government under Louis Philippe and the Whigs were ushered into office in Great Britain that November, it became impossible to deny that the Great Powers were deeply divided on the questions of revolution and domestic political legitimacy—a division that was codified in the Treaty of Münchengrätz and the Convention of Berlin. Both were signed in 1833, and a new Quadruple Alliance among the Western powers (this time including Spain and Portugal) signed the following year (Bulwer, 1871, 180–181; Bourne, 1970, 224).

Despite these rifts, the Concert system and the system of cooperation that it institutionalized survived for another two decades.[4] Although consultation took the form of ambassadorial conferences rather than Congresses, the Concert system accommodated the ideological differences of the Great Powers and helped to moderate clashes between their contradictory foreign policies. In the case of the Greek war of independence against the Ottoman Empire, for example, the British position shifted increasingly toward sympathy for the Greeks throughout the 1820s, and despite their antipathy toward nationalist uprisings Tsar Nicholas and Charles X joined the British to establish Greek independence in the London Protocol of 1830.[5] Even the revolutionary uprisings of 1848 across the Continent, which left only Russia and the United Kingdom untouched, and the subsequent wave of counterrevolution were not sufficient to destroy the Concert: In 1852, the Great Powers met again in London to resolve the issue of the Danish succession as well as a Danish dispute with Schleswig-Holstein.[6] The Concert only truly failed once, albeit spectacularly, when Russia, the Ottoman Empire, France, and Great Britain stumbled into a pointless and deadly war in the Crimea that left the Concert in tatters.[7]

The success of the Concert was due in no small part to its ability to resolve issues that might otherwise have caused conflict. Having just survived a twenty-five-year bloodletting, sparked by the French Revolution, that rivaled the Thirty Years' War in lethality, the Great Powers were initially unanimous in their commitment to the goal of suppressing and deterring national uprisings aimed at constitutional government. Even as that unanimity faded, the rules and norms of the Concert system remained intact: While the Great Powers disagreed over issues such as the independence of Greece in the 1820s and Belgium in the 1830s, they continued to manage these issues collectively and in a manner designed to avoid conflict. This cooperation is especially noteworthy given that the liberalization of the Western powers during this period actually created a *new* cause of war—humanitarian intervention—that by all rights should have been a major source of friction with the Eastern monarchies.[8]

That continued cooperation was likely the result, at least in part, of the continued ability of the Great Powers to resolve commitment problems that might otherwise have threatened the peace. Mutual restraint for mutual benefit was inherent in the idea of the Concert from the onset and helped to moderate concerns about the severity of the worst-case scenarios that might result from conflict. The act of meeting itself helped to ameliorate commitment problems: Discussion in a public forum forced the participants to justify their positions in terms of the collective good, thereby limiting their own range of options.[9] Because each knew that all had done so and would likely continue to do so, these meetings helped to reduce their concerns about worst-case scenarios.

At the same time, one of the Great Powers' first acts following the war was to codify a territorial settlement that included independent small states in the center of Europe, an arrangement that provided a buffer between the Great Powers and ensured that any attempt to change it would provoke a coalition sufficiently strong to deter the revisionist state. Those territorial circumstances, in turn, made possible a continuing Concert arrangement based on mutual restraint.[10]

The resolution of information problems, by contrast, seems to have been a less important driver of peace in the Concert, though far from irrelevant. As I noted earlier, the Concert was barely an institution itself, and it spawned no formal subordinate institutions to help manage conflict. The Concert was primarily designed to reduce conflict among the Great Powers and was not reluctant to use force toward that end, so no security was gained by a reduction in concerns about the behavior of subordinate states. That said, the Great Powers did establish and enforce rules of international conduct, and those rules did help to shape the expectations of smaller states and subnational groups about the conditions that would and wouldn't lead to intervention and war. The fact that those expectations played a nontrivial role in deterring revolution became evident when they were called into question—as, for example, when the success of the July Revolution of 1830 in France led to unrest in Belgium, Poland, and Italy. The importance of the Great Powers' likely reactions is illustrated by the fact that, in August, Belgian revolutionaries went so far as to send representatives to London and Paris to sound out the diplomats there (Artz, 1934, 273).

In sum, Great Power politics during the first half of the nineteenth century benefited from a perfect storm of compatible if not identical interests, well-designed political geography, and a durable international order founded on both. Over the course of almost forty years, the interests of the Great Powers changed as the societies of the Western powers evolved, but the rest of the Vienna settlement was solid enough to survive the transition and continued to provide an international order within which peace thrived.

The Mid-Nineteenth Century

The late 1840s and early 1850s were a period of momentous change in Europe. Austrian Chancellor Klemens von Metternich, who had worked tirelessly and brilliantly to preserve the territorial integrity of Austria since 1815, was driven from office in 1848. After a series of ephemeral successors—one of whom, Anton Freiherr von Doblhoff-Dier, lasted only ten days—he was eventually replaced by Prince Felix von Schwarzenberg. In the same year Louis Napoleon, a proponent of nationalism and constitutional government, rose to power as president of the French Second Republic, only to stage an *autogolpe* in 1852, establish the Second Empire, and take power as Emperor Napoleon III. The year 1852 also saw the rise of a moderate liberal, Camillo Benso, Count Cavour, to the prime ministership of the northern Italian Kingdom of Sardinia. The Treaty of Paris, which brought the Crimean War (1853–1856) to a close, rendered the Black Sea neutral territory and forbade fortifications in the Crimea, removing Russia from the ranks of the status quo powers. Finally, Tsar Nicholas I of Russia, who had taken up the defense of the monarchical principle against the threat of revolution with a grim consistency of purpose that his predecessor, Alexander I, wholly lacked, died in 1855 and in so doing removed the biggest obstacle to revolution in the East. His successor, Alexander II, is best known for (relatively) liberal reforms, including emancipation of the serfs. His foreign policy was markedly more moderate and pragmatic than that of his late father.

As a result, by the mid-1850s the remaining Great Power consensus on how to manage continental security had effectively evaporated. In practical terms, the Eastern monarchies' commitment to prevent revolution had become a dead letter. More generally, though, the old Concert had been grounded in a conservative vision of social order that was increasingly seen as oppressive and outdated (Schroeder, 1994, 803). The rise of "a new generation of young men in a hurry"—von Schwartzenberg, Louis Napoleon, Cavour, and before long Otto von Bismarck—"who were ambitious for their own countries and no longer willing to abide by the collaborative principles and practices invented and followed by the statesmen of Vienna" (Craig and George, 1995, 31) effectively doomed the international order that had been conceived at Vienna.

The implications of the fading of the Concert were soon felt in Italy. The Italian city-states had a long history of independence to varying degrees, succumbing to domination first by the Spanish in the mid-1500s and later to the Austrians in the 1700s. During the Napoleonic Wars Napoleon created a reorganized Kingdom of Italy with Napoleon himself as King, but under the terms of the Vienna settlement Italy reverted to Austrian control. The Napoleonic Wars nevertheless rekindled the idea of Italian independence—an idea that was consistently suppressed by the European Concert. Italy's most sustained attempt at breaking away from Austria came during the revolutions of 1848 in the First Italian War of Independence, which as the name implies did not produce the intended result.

The changed circumstances of the 1850s, however, produced a very different outcome, in large part because the Italians had gained a Great Power sponsor. In 1858, Napoleon III and Cavour met at Plombières-les-Bains in eastern France to hammer out the broad terms of an agreement that would result in France supporting Sardinia in a war against Austria with the goal of producing a northern Italian confederation under the control of Sardinia. A formal agreement was concluded in 1859, along with a separate agreement to forestall Russian involvement with the promise of support for a revision of the post-Crimea territorial settlement.

The story of how the Plombières plan resulted in a greatly expanded Kingdom of Sardinia that would in short order form the nucleus of an independent Italy is long and, although entertaining at times, mostly irrelevant. The upshot is that, by 1860, Sardinia had incorporated Tuscany, Parma, Modena, and the Papal States and had compensated France via the cession of Savoy and Nice. These changes to the status quo—the territorial expansion of France and the expansion of Italian influence at the expense of Austria—had been seen as two of the biggest threats to the peace when the Concert was established. Russia never interceded to prevent the erosion of the monarchical principle, nor did they or the British act to prevent French expansion. It's true that Prussia was willing to aid Austria, but only in a defensive capacity, and in a significant blunder the Austrians forfeited their help by taking the offensive against Sardinia. In short, the war confirmed that the Great Powers' commitment to act in the interests of peace in Europe had degraded to the point that it was no longer effective or even meaningful.

In the absence of such a shared commitment, three more wars followed in rapid succession: the Second Schleswig War of 1864, the Austro-Prussian War of 1866, and the Franco-Prussian War of 1870.[11] By the time these wars came to an end, they had completely overturned the territorial settlement that had been carefully designed at Vienna to preserve the peace. All of the buffer states in central Europe had been absorbed. The presence of a strong country, Germany, at the center of the Continent completely transformed the security calculus for European leaders. Moreover, two standing rivalries—between Russia and Austria-Hungary in the Balkans, and between Germany and France as a result of the German seizure of Alsace-Lorraine—represented a constant threat to the general peace of Europe.

The Bismarckian System

Europe in the 1870s therefore found itself in a situation too dangerous for comfort but not dangerous enough to compel concerted action to forestall that danger.[12] The Concert consensus had been destroyed and the territorial settlement upset, but the Great Powers lacked the agreement on principles and shared perception of threat necessary to forge a new international order.

Fortunately, German Chancellor Otto von Bismarck had both the ability and the motivation to construct a new, limited international order designed to keep the peace. Without a doubt Bismarck, by both unifying Germany and alienating France, had created a significant part of the problem that he now sought to resolve. Because France could be counted on to join any anti-German coalition, Germany faced the very real prospect of encirclement. The most obvious way to forestall that threat was to build up Germany's military capabilities, but a Germany that couldn't be threatened by its neighbors would itself become a major threat—one that would probably attract the opposition of the British.

Bismarck was no Metternich. He had no designs on constructing a security structure that would maintain the peace of Europe.[13] In order to resolve Germany's security problem, however, he was essentially forced to provide international order by initiating a comprehensive and unorthodox series of alliances with the other Continental powers (Langer, 1966, 195). He formulated his strategy succinctly: "All politics reduces itself to this formula: Try to be *à trois* as long as the world is governed by the unstable equilibrium of five Great Powers" (Joll, 1992, 45). The obvious part of the strategy is remaining *à trois;* the less obvious but more critical part is the unstable equilibrium. As we will see, Bismarck's treaties often worked to forestall coalitions rather than to form them.[14]

His first step was the signing of a military convention between Russia and Germany in 1873, each state pledging to come to the other's aid in case of attack with a force of two hundred thousand men.[15] Austria-Hungary promptly joined the two in a weaker agreement, the first *Dreikaiserbund.* The union has occasionally been compared to the Holy Alliance after Troppau and the Berlin Convention, but the two are dissimilar in the extreme: While the Berlin Convention implied an unlimited commitment to the cause of the monarchical principle, the 1873 accord promised only consultation in the event of a dispute. Its practical significance was limited to the Balkans, an area that threatened to drag Austria-Hungary and Russia into conflict.

The *Bund* was put to the test almost immediately by a crisis in the Near East and proved to be of little value. A Bosnian revolt in 1875 soon spread to Bulgaria and drew in Serbia and Montenegro. Russia, having bought off Austria-Hungary with the promise of Bosnia and Herzegovina, declared war on the Turks in 1877. The resulting settlement, codified in the Treaty of San Stefano, established substantial Russian gains in the Balkans and upset the delicate balance upon which the *Dreikaiserbund* had been predicated.

Bismarck was presented with a blank slate. Had he been able to count on an absence of alliances among his neighbors, he might have been able to pursue a less internationalist policy, but he could not. The scales of the balance were simply too delicate. An agreement between Germany and Austria-Hungary regarding the disposition of part of northern Schleswig in 1878 aroused Russian fears of collusion between the two and brought about discussion of the possibility of an eventual alignment with France. These results concerned Bismarck both because they suggested that Russia was an uncertain ally and because they brought his recurring nightmare of Franco-Russian encirclement one step closer to realization (Langer, 1966, 172–174).

Over the next four years, therefore, he constructed a substantially more complex series of alliances that nevertheless retained the key features of their predecessors: They were essentially defensive in nature, they papered over cleavages rather than exacerbating them, and they continued to isolate France. The first of these was a purely defensive alliance with Austria-Hungary in 1879, an alliance made possible by a calculated yet minor rift between Russia and Germany.[16] The terms of the alliance were quite straightforward. If either partner was attacked by Russia, the other was committed to joining it, and if either partner was attacked by another power, the other was committed to benevolent neutrality. The alliance was a model attempt at non-provocative deterrence: Its existence was made known at once, but its exact contents were secret, and in the event of a likely conflict the Tsar was to be made aware of them.[17] The short-term effects of the alliance were to increase the stability of Europe by deterring Russian aggression, ensuring the independence of Austria-Hungary, and restraining Austro-Hungarian ambitions in the Balkans.[18]

The next major step in the construction of a European major-power order was the inclusion of Russia in 1881. This agreement was far from a revival of the Berlin agreement of 1833: Whereas the former envisioned universal intervention in the event of revolution, the new *Dreikaiserbund* was designed only to ensure the neutrality of any two of the states in the event of an attack on the third. Significantly, since the agreement did not nullify the 1879 treaty between Germany and Austria-Hungary, the two treaties were contradictory, in letter and in spirit, to the extent that they attempted to formalize relations between Russia and the two other powers (Albrecht-Carrié, 1958, 180–181).

This fundamental tension was not entirely lost on the Russians, though the specifics were probably worse than they imagined. The fact remained that Germany would stand by Austria-Hungary in the event of a war, and under those circumstances it was prudent to consider the possibility of another ally. France was the only possibility, given Britain's relative disinterest in Continental commitments. Moreover, relations between Russia and Austria-Hungary were deteriorating over Balkan disputes, and for this reason the Tsar was unwilling to maintain the

Dreikaiserbund. In order to forestall the possibility of a Franco-Russian alignment, therefore, Bismarck in 1887 concluded a "Reinsurance Treaty" with Russia that specified that each state would promise to remain neutral in any war in which the other state was involved. The exceptions were two: Russia was not obligated to remain neutral if Germany attacked France, and Germany was not obligated to remain neutral if Russia attacked Austria-Hungary (Thomson, 1978, 469).

The series of treaties among the three states is an excellent example of the manner in which Bismarck made Germany secure without provoking British concerns regarding German power. He accomplished this feat by constructing inherently *un*balanced coalitions. Germany maintained a secure relationship with Russia despite Russia's conflictual relationship with Germany's ally: *contra* the Arab Maxim, the enemy of her friend was *not* her enemy. A quantitative analysis of diplomatic relations among Austria-Hungary, Germany, and Russia confirms the impression that the prevailing tendency was toward imbalance:

> if some states... retain the balance in their individual relations over time or perhaps even move to increase it, other states counteract the individual moves toward balance by actions that either maintain or increase the amount of imbalance in the triad. There is therefore no general tendency toward structural balance in the [Austro-German-Russian] triad in the 1880s. In fact, at the end of the decade, this triad moves from perfect balance to a high degree of imbalance.[19]

It was precisely this imbalance that defused the possibility of a coalition among the three states. As a result, Great Britain felt no compulsion to provide a counterbalance.

Italy, too, was brought into Bismarck's alliance system. The vehicle for doing so was the Triple Alliance. Franco-Italian clashes in Tunisia brought home to Italy the dangers of diplomatic isolation in 1881, and she hastened to come to an agreement with the central powers in the following year. Germany and Austria-Hungary promised to aid Italy in the event of a French attack; in the event of a war between one of the three and another of the Great Powers the other two were pledged at least to maintain benevolent neutrality; and if one of them was attacked and found itself at war with two of the other Great Powers, its allies were pledged to support it (Albrecht-Carrié, 1958, 180–181).

Only Britain refused to be a part of the Bismarckian system. The only agreements in which Britain did take part were temporary, limited, and secret, and even so they implied a very limited commitment. In 1887 Lord Salisbury was induced to support, in intentionally vague terms, two agreements with Italy and Austria-Hungary. The first agreement pertained to the maintenance of the status quo in the Mediterranean and surrounding seas; it therefore came to be called the first Mediterranean agreement. The second Mediterranean agreement extended the first to include the territories of the Near East. Neither promised anything more than consultation in the event of a crisis (Joll, 1992, 49). Neither was known, even in Britain, outside of the relevant committees until more than thirty years had passed, and when Germany pressed for renewal five years later the new British Foreign Secretary, Rosebery, explicitly disavowed them (Ensor, 1936, 198–199; Sontag, 1938, 277). The British had specifically refused association with the Continental alliance in 1879 and had again refused to play a larger role at the time of the Mediterranean agreements. When Bismarck explicitly offered to bring Britain into the Continental system with an alliance against France in 1889, the British again refused his offer (Ensor, 1936, 199). Bismarck was hoisted with his own petard: Faced with the task of ensuring German security without arousing British opposition, he had succeeded by unbalancing and entangling the European power structure in such a way that Britain had no reason to support him either.

In all, the systems of treaties established by Bismarck stabilized Europe neatly without provoking British opposition. Two characteristics were key. First, they were defensive in nature.[20] None of the states in question could count on support from its allies if it were to initiate hostilities. The members of the Triple Alliance had even gone so far as to announce that the alliance could not "in any case be regarded as being directed against England" (Albrecht-Carrié, 1958, 185). Second, they were riddled with contradictions. The 1879 alliance was nestled inside the *Dreikaiserbund* like a bomb wired to go off if the machinery of the latter were set in motion by the Russians to cover aggressive maneuvering in the Balkans. The *Dreikaiserbund* brought Russia into alignment with Germany and Austria-Hungary, while the Triple Alliance was based on at least the likelihood of Russian hostility. The intended effect was not mobilization and aggression but paralysis.

Bismarck's systems of alliances mostly didn't resolve commitment problems, because the alliances themselves represent exactly the sort of commitment that's impossible to arrive at in the presence of commitment problems. Indeed, he often sought provisions—a defensive alliance against France in the 1879 treaty, for example—that his allies were unwilling to provide (Eyck, 1964, 264–265). The main exception is a subtle one: France and Germany were unable to commit credibly to a long-term resolution to the question of Alsace and Lorraine, but Bismarck's skillful isolation of France deprived the French of the Continental ally that they would have needed to force the issue.

The relationship of Bismarckian diplomacy to information problems, on the other hand, is complex and fascinating. Bismarck's alliances conveyed important and credible information about Germany's intentions in the event of conflict, and they also bound the other Continental powers to conditional courses of action, the knowledge of which was valuable information both to Germany and to other states. To this extent, Bismarck's defensive alliances kept the peace by resolving information problems in a way that did not promote aggression.

At the same time, Bismarck also kept the peace by *hiding* information. While Bismarck's penchant for secrecy can be overblown—historian A. J. P. Taylor (1954, 266, fn. 1) rightly huffs that "Much nonsense is talked about the 'secrecy' of various alliances"—it is nevertheless true that some had secret provisions and others, like the Reinsurance Treaty, were entirely secret to everyone but the signatories for many years. Professor Edward Crankshaw's (1981, 369) characterization seems most apt:

> Bismarckian diplomacy might be defined as secret agreements exacerbated by leaks. Nothing could have been better calculated to keep everyone guessing and uneasy. That is to say, he allowed others to see that something was going on, but gave them no idea what it was.

While the Russians were immediately made aware of Germany's 1879 alliance with Austria, for example, the treaty that the three Great Powers signed two years later conveniently omitted mention of the fact that the 1879 treaty committed Germany and Austria to mutual defense in the event of an attack by Russia. The Reinsurance Treaty, concluded in 1887 and kept wholly secret until 1896, significantly undermined Russia's defensive commitment to Austria.[21]

While these treaties did reduce the other Great Powers' uncertainty about Germany's likely policy in the event of war, therefore, the information that they provided was neither complete nor wholly accurate. It was, however, incomplete and inaccurate in ways that prevented conflict. Bismarck's system was an improvised and inconsistent patchwork of commitments that he revealed only selectively. It succeeded by reassuring his allies—whether or not that reassurance was really warranted.

Weltpolitik

When Kaiser Wilhelm I died in 1888 and his only son, who suffered from an advanced case of throat cancer, followed him ninety-nine days later, Wilhelm's grandson, Friedrich Wilhelm, took the throne as Wilhelm II at age twenty-nine. The headstrong young Kaiser and his equally headstrong old chancellor clashed early and often, and in 1890, following an exchange of letters whose baroque formality did little to hide their venom, Bismarck was sent into retirement.

It is not clear how long a united Germany with a thriving economy, situated at the heart of Europe, could really have avoided posing a threat to the other Great Powers, the British in particular.[22] What is clear is that Kaiser Wilhelm's policies did nothing to forestall this outcome and quite a bit to accelerate it.

Bismarck's disappearance from the diplomatic scene left German foreign policy in the hands of a small group of men, none of whom managed to dominate it entirely. Kaiser Wilhelm held the most sway when he chose to exert it, but despite his eagerness he was inexperienced and prone to inconsistencies. Baron von Holstein, Bismarck's subordinate for more than a decade in the foreign office, remained entrenched in an obscure position but nevertheless had more say in the formulation of German foreign policy than anyone save Wilhelm and the chancellor. The three men who succeeded Bismarck—General Leo von Caprivi (who had once asked, "What kind of a jackass would dare to be Bismarck's successor?" [Massie, 1991, 111]),

Prince Chlodwig zu Hohenlohe-Schillingsfürst, and Prince Bernard von Bülow—varied in their experience in foreign affairs, from none to little (Dawson, 1919, 282–283; Craig, 1965, 23–24; Albrecht-Carrié, 1958, 207; Holborn, 1969, 309).

Caprivi's first move was to permit the Reinsurance Treaty with Russia to lapse in June 1890, an act that inadvertently affirmed his assessment of Bismarck's successor. Holstein differed substantially with his chief on the question of the Russian treaty, and he convinced Caprivi to call a meeting to discuss the question of the treaty's renewal on March 23 of that year. At that meeting, Holstein pointed out that Germany had much to lose by renewing the Reinsurance Treaty, given its inconsistencies with Germany's other obligations (in particular to Austria-Hungary), and little to gain, since nothing in the treaty formally prevented a Franco-Russian alliance. Moreover, the German promise to close the Straits in the event of war would alienate the British if it were revealed. Holstein convinced Caprivi, who in turn convinced William II, and the treaty was allowed to lapse despite substantial Russian efforts to make its burdens less onerous (Langer, 1966, 500–502).

Having "cut the wire" to St. Petersburg, the Germans proceeded to reaffirm the strength of the Triple Alliance, which was renewed in the form of a single treaty in May 1891. Italy in particular attempted to bring Great Britain into the agreement; Caprivi, realizing that the British would not be committed to participating in a Continental war, convinced the Italians to settle for a clause reaffirming the status quo in North Africa. Britain refused to undertake even this commitment and remained outside of the Triple Alliance. During the course of the negotiations the Italian Prime Minister, Rudinì, claimed that Great Britain had agreed to join the alliance; although false, the statement aroused no small concern in France and Russia (Taylor, 1954, 332–333).

While it was true that no clause in the Reinsurance Treaty had prevented Russia from forming an alliance with France, Bismarck had anticipated that the treaty itself would perform that function by substituting German friendship for French. The benefits of German friendship in Balkan affairs were substantial; alliance with France was a second-best option. It was, nevertheless, the only option remaining to the Russians following the lapse of the Reinsurance Treaty, and the circumstances surrounding the renewal of the Triple Alliance made both powers substantially more nervous. In August 1891 the two powers negotiated an exchange of letters promising consultation in the event of a threat to the peace and some form of agreement in the event that one of the two was actually threatened. A year later, the understanding became a formal alliance.[23]

The ratification of a Franco-Russian Alliance was the realization of Bismarck's worst nightmare. If France were attacked either by Germany or by Italy with Germany's support, Russia pledged to attack Germany with all of her available forces. If Russia were attacked either by Germany or by Austria-Hungary with German support, France pledged to do the same. In either case, the attack was to be carried out with the greatest possible speed, so that Germany would be forced to fight a war on two fronts. Although the treaty itself was secret, the visit of a Russian naval squadron to Toulon in the summer of 1893 spoke volumes. In this way, an unstable association of four Great Powers and an isolated fifth were transformed into two rival coalitions.

Germany's decision to pursue a more wide-ranging and aggressive "world policy," or *Weltpolitik,* made an already threatening security situation even more dangerous. In the 1870s and 1880s, few if any German politicians had aspirations for their country to be a world power. By the early years of the new century, most had come to believe that playing such a role was essential (Dawson, 1919, 396). It became common to yearn openly for "political domination and hegemony" and to make such assertions as, "Germany must achieve absolute hegemony in central and western Europe"; "Germany will be a world power or nothing"; and "If Germany does not rule the world... it will disappear from the map."[24]

The first concrete signs of a more aggressive German foreign policy came in 1897, when Admiral Tirpitz was made State Secretary of the Reich Navy Office. During the previous year, Tirpitz had toured China in search of a possible military and economic base. He chose the port of Kiaochow (Jiaozhou) on China's Yellow Sea coast. In August of 1897, two months after Tirpitz's rise to office, two German missionaries were "opportunely murdered"[25] nearby. The result was the German seizure, and subsequent ninety-nine-year lease under duress, of Kiaochow in January 1898. The lease was justified by reference to Germany's need for a coaling station for its Navy. The German Navy was as yet insubstantial, but now that a coaling station had been obtained, it became an absolute necessity, if for no other reason than to protect the coaling station. The first Navy Law was passed in April 1898.[26]

The German Navy Law of 1898 was a subtle piece of legislation, for the simple reason that the measure had to be approved by the same body (the Parliament, or *Reichstag*) which it intended to deprive of financial control of the Navy. Tirpitz could not, therefore, simply requisition all of the ships he desired; rather, he asked for a limited number while adding clauses establishing both a steady tempo of shipbuilding and automatic replacement of ships after a given number of years. Once these measures had been granted, they could not be revoked by the *Reichstag,* but until they had been granted, proposals for naval budgets had to be moderate and couched in mild terms.[27]

Due largely to Tirpitz's cautious gradualism, the British did not immediately recognize the German naval buildup as a substantial threat. In fact, to Britons at the time the most significant result of Germany's seizure of Kiaochow was a subsequent scramble for bases in China, which led to yet another Anglo-Russian conflict. At the same time, British colonial expansion from southern Africa and French expansion from the west collided in the eastern African town of Fashoda, which would become the site of a major conflict between the two powers by early autumn. The situation with Russia especially prompted another British approach to Germany.

The Navy Law of 1898 had envisioned a German fleet roughly half the size of the Royal Navy in the long run, a prospect which the British found not too unsettling. The 1900 bill envisioned a two-to-three ratio, however, and its passage occasioned substantial concern in London (Craig, 1978, 312). There was one more attempt to discuss an Anglo-German alliance, in 1901; again, the British found the Germans to be less cooperative than they might have hoped because the liabilities of a partial alignment with Great Britain (especially the risk of alienating Russia further) would outweigh the benefits. Bülow replied that cooperation would have to take the form of British membership in an expanded Triple Alliance, a price which Britain was still unwilling to pay (Taylor, 1954, 396–397).

This point marks the beginning of the Anglo-German rivalry and consequent European polarization which led to World War I. The fundamental tension was not hard to see: German *Weltpolitik* compelled growth and expansion, and British belief in the balance of power compelled opposition. The first sign of the impending clash was a quantitative and qualitative Anglo-German naval arms race. In 1902, Great Britain formally adopted a three-power standard. The Germans responded with a new Navy supplement in 1905, to which the British responded with the introduction of the *Dreadnought*-class battleship in 1906. Additional German supplements in 1906 and 1908 brought another increase in Britain in 1909. British warship tonnage more than doubled between 1900 and 1910 and had nearly tripled by 1914; German tonnage more than tripled between 1900 and 1910 and had nearly quintupled by 1914.[28]

At the same time, the polarization of the Continent increased, and Great Britain was drawn farther and farther into the Franco-Russian camp. The first major event in this regard was the *entente cordiale* between England and France in 1904, essentially a settlement of outstanding colonial issues. A German attempt to disrupt the coalition which was forming against it failed when a defensive alliance between Germany and Russia, signed in 1905 by William II and Nicholas II at the former's insistence, was reconsidered and then rejected by the latter upon greater reflection. The *entente* both alienated Germany from France and brought England closer (Joll, 1992, 80). The first of two crises between Germany and France over Morocco led the British to contemplate intervention on the European continent for the first time since 1864. This fact, as much as the increases in the British Navy, was a decisive turning point for the British:

> Once the British envisioned entering a continental war, however remotely, they were bound to treat the independence of France, not the future of Morocco, as the determining factor. The European Balance of Power, which had been ignored for forty years,

—or, more accurately, which had not required British attention for forty years—

> again dominated British foreign policy; and henceforth every German move was interpreted as a bid for continental hegemony. (Taylor, 1954, 438)

As a result, Great Britain and France began to coordinate their military plans on the Continent in 1906 (Herrmann, 1996, 55). Another *entente* between Great Britain and Russia followed in 1907. A crisis over Serbia in 1909 led to a further tightening of the alliance between Germany and Austria-Hungary: Germany not only issued an abrupt ultimatum to the Russians but made it clear to Austria-Hungary that it could expect German support if it found an invasion of Serbia to be necessary. The defensive treaty of 1879 had become an offensive one (Craig, 1978, 323). The second Moroccan Crisis between Germany and France in 1911 drove Britain and France further into one another's arms, and by 1912 the two had agreed on a naval division of labor as well (Thomson, 1978, 535). By 1914, the Continental coalitions were so tight that when Gavrilo Princip succeeded in assassinating Archduke Franz Ferdinand in Sarajevo the war spread to the entire continent. The Germans, having succumbed to the "cult of the offensive" (Van Evera, 1985) and not really believing in the neutrality of Belgium,[29] invaded the latter, ensuring that Great Britain would join the war as well.

The rise of German power in the early 1900s created a classic example of a commitment problem: a situation in which rapid changes in relative power prevent a rising state from being able to make a credible commitment to peace. A famous memo by Sir Eyre Crowe of the Foreign Office, dated January 1, 1907, captures this problem neatly. The question of whether Germany was "aiming at a general political hegemony and maritime ascendency, threatening the independence of her neighbours and ultimately the existence of England" or was merely "seeking to promote her foreign commerce, spread the benefits of German culture, extend the scope of her national energies, and create fresh German interests all over the world," Crowe wrote, was irrelevant. The growth of the German Navy supported both hypotheses, but, more importantly, even if Germany were pursuing the second path, it could revert to the first at any time. Its capabilities relative to those of its potential opponents therefore constituted a de facto threat to Great

Britain (Bourne, 1970, 489–490). Less obviously, other countries' military responses to Germany's rising power created another commitment problem: Rapid shifts in relative military power created a perception on the part of Germany that they faced a closing "window of opportunity" to engage in a war that they might reasonably expect to win.[30]

At the same time, the requirements of anticipating the reactions of so many other Great Powers produced an information environment that was exceptionally conducive to miscalculation and war.[31] The Kaiser mistakenly believed that his cousin, the Tsar, would be so horrified by regicide that he would refuse to back the Serbs in the summer of 1914, and both the Kaiser and his Chancellor, Theobald von Bethmann-Hollweg, were confident that, even if the Russians were to intervene, the British most likely would not.[32] The importance of Germany's miscalculation became evident when it became clear that the British were weighing intervention. Bethmann-Hollweg was so alarmed at the prospect that he proposed a variety of concessions, including *signing the entire German fleet over to the British,* to prevent them from becoming involved.[33]

It is impossible, of course, to re-run history and determine whether a Concert-style or Bismarckian international order would have prevented World War I, but on the surface at least the proposition seems plausible. Bismarck well understood the importance of the isolation of France, and even if he hadn't been able to maintain it for another twenty-four years, one suspects that he would have come up with some way to counterbalance it. Certainly, the polarization of Europe into two opposing camps, the Triple Alliance and the Triple Entente, would have been anathema to him. Had the Concert of Europe, or some more modern analog, been functioning at the time of the July Crisis there would certainly have been a multilateral conference to try to resolve the Austro-Serbian dispute. Sir Edward Grey, the British Foreign Secretary, did propose such a conference, although he did so very late in the game. But the responses of the other Powers—Russia preferred direct talks with the Austrians, Germany sought to undermine those talks, and Austria contemplated accelerating its declaration of war in order to prevent mediation[34]—is indicative of the extent to which they had simply given up on the idea of collective management of European security.

The Interwar Period

Just as the Napoleonic Wars had done, World War I focused the attention of the Great Powers on measures that would prevent another major conflict. The results of that focus were far more substantial in 1919 than they had been in 1815, however. The liberal critique of Europe's deteriorating conservative order had grown louder as the war wore on, and by the time the armistice was signed in 1918, plans were in place for a comprehensive, liberal world order to take its place.

The order that was engineered by the statesmen of 1919 and fleshed out throughout the 1920s was staggering both in scope and ambition. Its centerpiece, the League of Nations, set as its primary goal the promotion of world peace. It sought to do so both by preventing wars when they threatened, via collective security and the settlement of international disputes, and by promoting international conditions that were thought to make war less likely—social justice, decreases in armament levels, the protection of minorities, the administration of former colonies, and so on.

History has not been kind to the League. Its provisions, in retrospect, seem hopelessly naive: Member states were expected to come to one another's aid in the event of aggression, an obligation that deftly elided the problem of agreeing on who was the aggressor in the first place. Members were also expected to submit disputes to a Permanent Court of International Justice rather than go to war to resolve them, on the questionable assumption that such disputes were amenable to negotiated settlement. Armaments races having been seen as a major cause of the preceding war, states were also expected to disarm as much as possible within the limits dictated by considerations of national security.

Despite the burden of unreasonable expectations, the League did chalk up some very notable successes in the 1920s. It mediated territorial disputes in places like Upper Silesia and the Saarland, preserved Albania against encroachment by Greek and Yugoslav troops, and sent a small peacekeeping force to resolve a clash along the border between Colombia and Peru. It showed its ability to defuse conflicts in Bulgaria and Liberia. The spirit of the League was also reflected in the conduct of international relations more generally. While disarmament conferences had not been unheard of prior to World War I—the Rush-Bagot Treaty of 1817 led to the demilitarization of the Great Lakes, for example—the Washington Naval Treaty of 1922 represented a profound victory for the world disarmament movement. The Geneva Protocol of 1925 sought to limit the horrors of war by outlawing poisonous gases and biological warfare. The Locarno Treaties of 1925, in securing postwar borders and normalizing relations with Germany, ushered in the "Spirit of Locarno," a sense of optimism that war had indeed been banished. This sentiment found its ultimate expression in the growing "outlawry of war" movement and, finally, in the Kellogg-Briand Pact of 1928, in which signatories renounced war as an instrument of policy.

Unfortunately, its failures were more notable. The World Disarmament Conference of 1932 foundered on the problem of French and German armaments and gave Hitler an excuse to withdraw from the League. The Japanese invasion and occupation of the Chinese region of Manchuria, a flagrant act of aggression, was labeled as such by the League Assembly, which demanded Manchuria's return to China. Rather than comply, Japan too left the League. Finally, an Italian invasion of Ethiopia in 1935 generated only condemnation and ineffectual sanctions on the part of the Western powers.

In the broader context of the history of international affairs, the League was a substantial landmark. It was vastly more ambitious and organized than the old Concert of Europe. In the words of George Scott (1973, 15), the League Covenant

> stands today, no less than it did in 1919, as a testament to the aspiration of man to govern his affairs by reason, to assert the concept of justice into the settlement of international disputes and to enshrine the collective interest of all nations as supreme above the interests of any group of nations or of any individual nation.

Yet it is not without reason that the League is more commonly associated with an impractical Utopianism. It had no autonomous forces to command. It was founded on the principle that the nations of the world so overwhelmingly rejected aggression that they would be willing to pay the costs of standing up to it whenever it threatened, despite the absence of any formal obligation to do so.[35] And it depended heavily on the assumption that states would follow international court rulings just as individuals comply with the rulings of domestic courts, despite the crucial difference that the former were in the last resort unenforceable. It was, to put it more succinctly, built on a surfeit of hope.

That hope might have been at least somewhat more justified had it not been for a second important feature of the international system in the interwar period: The emerging Western liberal order embodied by the League faced opposition from three other attempts by Great Powers to create international order that both undermined its universal aspirations and fed on its apparent weakness during the Great Depression. The result was a fundamental clash among what Barry Buzan (2017, 238) calls the "the four ideologies of progress" of the nineteenth and twentieth and 20th centuries, "liberalism, nationalism, socialism, and 'scientific' racism," each of which in some measure contributed to an international order promoted by one or more of the period's most powerful states.

The first of these, the Soviet attempt to ignite a worldwide Communist revolution, was decidedly unsuccessful beyond the Soviet Union itself. Save for Béla Kun's short-lived but vicious Hungarian Soviet Republic of 1919, the Soviet attempt to free the workers of the world from their capitalist shackles failed utterly, and the Soviets soon focused on the development of what they called "socialism in one country." Even so, the success of the Bolsheviks in overthrowing the provisional government in 1917 inspired a brief but intense "red scare" in the United States, and the Soviet-sponsored Communist International held world congresses throughout the 1920s and 1930s as a means of coordinating revolutionary pressure on the international bourgeoisie.[36]

The second attempt at an international order began with an effort to liberate and protect Asian countries from European control and evolved into a regional Japanese empire that came to be known as the Greater East Asia Co-Prosperity Sphere. Here, a potent admixture of nationalism, militarism, and racism drove both expansionism and an inherent hostility to the Western liberal order.[37]

The third attempt at an international order, of course, was Nazi Germany's bid to create a Third Reich that would span the globe. Sir Michael Howard nicely captured the essence of Nazi goals:

> In so far as Hitler had a world vision it was one of a hegemony kept on its toes by continual conflict, a caricature of a Roman empire in

> which the rest of the world consisted either of subordinated associates or barbarian enemies. Peace did not exist in the fascist vocabulary, except as a term of mockery or abuse. . . . For [Hitler], war was not simply an instrument for creating a new order: it was the new order.[38]

The view, advanced by Professor Mueller (2007, 2), that Hitler's Germany was sufficient in and of itself to explain World War II is probably not far from correct. War was integral to Nazi Germany's foreign policy, a fact that set it on an inevitable collision course with the West. It does not follow, however, that a world free of the Nazis would have been a world free of conflict. Japan's regional ambitions and the Soviet Union's global ones were fundamentally at odds with the Western liberal order that proliferated in the 1920s. The cooperation of the Soviets with the West during World War II only proved that, if politics make strange bedfellows, wars make even stranger ones. Only a stark imbalance of economic and military power and, before long, the rise of a common enemy had prevented a more active contest between the Soviets and the West, and such a contest appeared anyway once the war was over, despite the devastation that had been inflicted on the Soviets.

The League is widely seen as the international order that failed, and it's hard to argue with that assessment. It does, however, deserve to be qualified. When the League worked, it worked well. Its members' commitments to nonviolent dispute resolution and collective security did contribute to the resolution of commitment problems in a handful of disputes that might otherwise have led to war. Even into the late 1930s, surprisingly few conflicts among League members ever took place. That outcome, however, is mostly due to the fact that major-power aggressors like Germany and Japan simply left the League rather than face punishment. So the expectations laid out in the previous chapter—decreased conflict within international orders and increased conflict between them—turn out to be an accurate description of the interwar period, but partly for the wrong reason: In the 1930s, the League screened out potential aggressors rather than constraining their behavior.

The Cold War

After World War II, the major powers came together around an attempt to build a global order, in the form of the United Nations, that retained many of the best features of the League without the flaws that contributed to its demise. The United Nations Charter reflected a considerably broader consensus than the League Covenant had: Fifty countries ratified the former at its inauguration. Articles 39–51 of the Charter gave the United Nations, and especially the Security Council, both the authority and the means to maintain and restore peace in the event of a breach. And the founders of the United Nations went out of their way to ensure the continued participation of the world's most powerful countries.

That last accomplishment came at a cost, however. While the Covenant required unanimity in order to act, the Charter allowed for a more flexible voting scheme in which no single power had a veto—except for the five permanent members of the Security Council. That concession amounted to a frank recognition that the mandate of the United Nations did not extend to the vital interests of the Great Powers, however those powers chose to define them.

Although few realized it in 1945, the manners in which two of the Great Powers defined their interests were so diametrically opposed that they soon gave rise to two additional international orders, based in Washington, DC, and Moscow. American decision makers initially believed that the Soviet Union's foreign policy would be guided by pragmatic rather than ideological thinking.[39] The Soviets expected a falling-out among the Western democracies: As good Leninists, they believed that capitalism's excess production would lead to a constant search for new markets and conflict among the major capitalist states.[40] From the constitution of the postwar Polish government to the composition of governments in Bulgaria and Romania, however, it soon became apparent that the Soviet desire to ensure friendly governments within its sphere was not at all consistent with the Western desire to see democratic governance in those countries. By early 1947 the announcement of the Truman Doctrine and the Marshall Plan left little doubt that the former allies were fundamentally working at cross purposes. The creation of the Communist Information Bureau, or Cominform, a body designed to coordinate cooperation within the Communist world and form a solid front against the West, followed in October of that year (Zubok and Pleshakov, 1996, 54, 130–131) and the North Atlantic Treaty Organization (NATO) was formed only two years later.

As institutions—NATO, SEATO, the Warsaw Pact, the European Union, the Council for Mutual Economic Assistance (Comecon)—that reflected and reinforced different facets of the emerging orders began to form, it became clear that the postwar international order would be a fractured and fragmented one. The Western democracies came together behind the leadership of the United States, which alone possessed the requisite financial wherewithal both to build up Western military power and prevent the sort of economic distress that had fueled the rise of Communist and fascist movements in the interwar period. The Communist international order was not as unified as its Western counterpart, due in large part to the different priorities of the Soviets and the Chinese Communists.[41]

At the same time, the nascent global order that was initiated with the United Nations began to take more concrete form. The United Nations itself developed into a robust and functional organization with effective economic, social, and judicial organs that it uses to provide humanitarian assistance and promote development and human rights, but the Security Council's monopoly on the authorization of the use of force allowed East and West to minimize its role within their own spheres of interest.[42] A variety of United Nations satellite organizations, independent intergovernmental organizations, treaties, and agreements—the International Monetary Fund and the World Bank, the Universal Declaration on Human Rights, the General Agreement on Tariffs and Trade, the Helsinki Accords and the Conference on Security and Cooperation in Europe, the Convention on the Elimination of all Forms of Discrimination Against Women and the Committee Against Torture, and many more—helped to define the political and economic dimensions of global order in the postwar world.

While the principals in the Cold War went to great lengths to avoid clashing directly—Soviet pilots participated in the Korean War, for example, but they wore Chinese uniforms, avoided the front lines, and did not speak Russian over the radio—each suspected and opposed encroachment by the other in the most remote corners of the world. We can't attribute *every* use of force during the Cold War to the competition between the liberal West and the Communist East, of course—some conflicts, like the Soccer War between El Salvador and Honduras in 1969, were entirely localized. But because international conflict in this period played out within the broader context of the East-West struggle over international order, political elites often sought assistance from one superpower or the other, and the superpowers were often all too willing to oblige.

As a result, the global clash of international orders fueled conflict between American and Soviet allies and proxies in locations as diverse as the Taiwan Straits, the Suez, Eritrea, Vietnam, Cambodia, the Korean peninsula, Indonesia, Algeria, and the Middle East. Adding civil wars fought by American and Soviet proxies would make the list much longer. Conflicts short of war, such as the U-2 incident, the Bay of Pigs invasion, the Entebbe raid, the downing of Korean Air Lines flight 007, and the American invasions of Grenada and Panama, were so ubiquitous that they came to be expected.

The United Nations, meanwhile, sought to reduce international conflict in a variety of direct and indirect ways. Unfortunately, it is extremely difficult to say how effective it was, for one simple reason: The United Nations doesn't try to solve the easy problems. The worse a conflict is and the more potential there is for widespread loss of life, the greater the odds that the United Nations would become involved in an attempt to lessen the damage.[43] Because the UN preferentially involves itself in conflicts where we *expect* the outcome to be worse, we can't simply compare the outcomes of conflicts to gauge the impact of UN involvement: Even if such involvement saves a lot of lives, cases in which the UN involves itself could still be considerably more intractable, on average, than cases in which it does not.[44]

Even so, scholars have uncovered some connections between intergovernmental organizations (IGOs) and conflict that warrant considerable optimism. The earliest of these studies argued that joint IGO membership reflected commitment to international law, one leg of the so-called Kantian tripod (democracy, trade, and international law) that produces peace (Russett, Oneal, and Davis, 1998; Russett and Oneal, 2001). More recently, scholars have shown that highly institutionalized IGOs provide information that helps to reduce the probability of conflict onset (Boehmer, Gartzke, and Nordstrom, 2004; Shannon, Morey, and Boehmke, 2010), while IGOs designed to mitigate commitment problems decrease the duration of conflict (Shannon, Morey, and Boehmke, 2010). Similarly, the network of ties that evolve from joint membership in IGOs appears to reduce conflict (Dorussen and Ward, 2008). While this is a comparatively young literature, the results are increasingly specific and on the whole encouraging.

The overall picture during the Cold War was, on the whole, fairly complex. Although the superpowers mostly went out of their way to avoid direct conflict, members of the so-called First World—the industrial, capitalist countries that were aligned with the United States against the Communist bloc—and those of the Communist Second World clashed around the globe, sometimes directly, but mostly in proxy wars in the third world. In some cases, conflict was inspired directly by the clash of international orders. In most, the backing of the superpowers gave the combatants more confidence than they otherwise would have had in their own capabilities. The uncertainty surrounding superpower support, moreover, raised the perpetual specter of commitment problems: In the event of a negotiated settlement, one or the other superpower could be counted on to try and undermine it. In this way, the involvement of the superpowers both decreased the stability of peace across orders and made the prospects for a lasting settlement more tenuous. At the same time, however, everyone enjoyed the conflict-suppressing benefits of membership in a growing network of intergovernmental organizations.

The Post–Cold War Period

If the Cold War showed us what can happen when two superpowers, each promoting a vision of world order that is antithetical to the other, fuel conflict across the globe, the post–Cold War period shows us what happens when one of those superpowers gives up the fight. Between 1989 and 1991, the Soviet Union renounced its previous attempts to spread its ideology overseas, allowed revolution to undermine Communist control of its East European empire, and in the end even gave up its own territorial integrity, fracturing into independent countries that had once been constituent republics of the USSR.

The Western liberal order, by contrast, expanded substantially in the post–Cold War world. NATO created the Partnership for Peace, a program for bilateral cooperation with nonmember states in Europe, and engaged in four rounds of expansion that resulted in NATO membership for thirteen new states. The European Union succeeded the European Economic Community in 1993 and soon expanded from twelve countries to twenty-eight.

At the same time, the global order of the post–Cold War world has, unsurprisingly, come to reflect the values and priorities of its Western liberal nucleus to an even greater degree than it previously had. While no formal peace treaty defined the post–Cold War world in the same way that the Vienna Final Act had defined post-Napoleonic Europe, the 1990 Charter of Paris for a New Europe sought to chart a course for the continent that emphasized democracy, human rights, economic openness, social justice, and environmental responsibility. The World Trade Organization (1995) and the International Criminal Court (1998), to take two prominent examples, greatly strengthened international law in the areas of trade and human rights. The increased breadth and depth of the dimensions of world order most central to Western liberalism, in turn, blurred the distinction between the global order and the expanded postwar liberal order at its core. As Professor G. John Ikenberry puts it, "[i]n critical respects, it was a one-world system in which the United States and the organizational logic of the Cold War–era Western order remained at its center" (Ikenberry, 2011, 236).

The effect of these changes on the rate at which force was used in the international system was striking. The median rate of the use of force during the Cold War was 15.4 uses of force per thousand politically relevant pairs of countries per year; at the end of the Cold War that rate dropped by two-thirds, to 5.6. Even the September 11, 2001, attacks in the United States and the subsequent wars in Iraq and Afghanistan have failed to produce a significant uptick in the rate of international conflict initiation. They *have* produced a noticeable increase in the number of initiations of extra-systemic wars, or wars between states and nonstate entities (see the discussion beginning on page 96), in the very recent past. It remains to be seen, however, whether that trend will continue.

From the perspective of international order, the present lull in conflict therefore owes at least as much to Mikhail Gorbachev as it does to the pacifism of advanced Western democracies. The Western liberal order plausibly explains the absence of conflict in the West. But the reformulation of the Soviet Communist world order and the subsequent dissolution of the Soviet Union deserve credit for eliminating conflict between two international orders that had, for the first time in history, fought one another across virtually every habitable piece of territory on the globe.

Data and Analysis

In addition to helping shed light on the mechanisms that drive sea changes in conflict over time, this brief historical survey is also useful for highlighting subsets of the militarized interstate dispute data that we can compare, as a check on the qualitative historical record. To carry out such an analysis, I first asked a research assistant to compile data on international orders from 1815 to the present so that I could sort the data by order and by period and compare rates of international conflict initiation across them.[45]

The question of how to measure international order is a challenging one, especially across such a long period and with such a diverse array of orders. The Western liberal order, for example, has given rise to a bewildering array of institutions, some of which are more related to order than others. The Concert of Europe exhibits the opposite pathology: Its lack of institutionalization makes it difficult to know when exactly it ended.

To the extent possible, we used the historical literature as our guide and made the most sensible decisions we could. Some decisions were fairly easy. We dated the Bismarckian period by looking at the rise and fall of Bismarck as chancellor of a united Germany, and there are clear records of membership in the League of Nations and the United Nations. Others were a bit more subjective: We ended the Concert of Europe at the Crimean War, for example, largely because I think that's the most reasonable point at which to end it, for reasons that I described above.

The most challenging decisions involved the Western liberal international order during and after the Cold War. Commentators have a pesky tendency to define it implicitly by asking which countries are mostly on board with the Western liberal project. The problem, of course, is that we want to understand the relationship between international order and conflict, so if we were to define membership in an international order by virtue of whether a country is challenging that order I would rightly be accused of circular reasoning. The closest thing to a formal operational definition that I have found is Professor David Lake's 2009, 68–76) index of security hierarchy, which is based on overseas troop deployments and shared alliances. As Lake himself notes, however, the measure is noisy: Some close US allies, like Israel, lack formal defense agreements, while the presence of troops stationed abroad, as they currently are in Iraq and Afghanistan, can be indicative of an unstable country that is "in play" rather than of an established member of the Western liberal order.[46]

In the end, we chose to include three categories of countries: NATO members, other developed countries that were aligned with the West (Australia, Japan, South Korea, etc.), and neutral developed countries that leaned toward the West (e.g., Ireland). Members of the Communist international order included Warsaw Pact countries as well as other Communist states. These decisions best captured the "first world–second world" distinction that was often used to sort countries into one camp or the other during the Cold War. After the end of the Cold War, we added to the Western liberal order those countries that joined either NATO or the European Union or both, starting in the first full year of membership.

I then calculated the rates of conflict initiation for each of these international orders as well as for pairs of countries that crossed the boundaries of an international order. Because, as I noted in chapters 1 and 3, there is an element of chance in international conflict, I calculated credible intervals to capture the uncertainty of each estimate, using the bootstrapping technique described in chapter 5.[47]

The results are displayed in Figure 8.1. The blue dots represent international orders; the red dots represent interactions among pairs of countries that are not in the same international order (or, in the case of the second and fourth periods in the nineteenth century, are not in any international order at all). The lines to either side of the dots span the credible interval of the estimate and in so doing illustrate its uncertainty.

In general, the data line up very nicely with the history of the last two hundred years that I've laid out above. The Concert period and the Bismarckian order are both characterized by very little conflict, while the other two periods that make up the pre–World War I era are considerably more conflictual. There was very little conflict among the countries within the League of Nations—though, as I noted above, this is partly a result of the fact that aggressive, revisionist countries dropped out of the League in the 1930s. By contrast, the relationship between League countries and the countries at the heart of the three competing interwar orders—Germany, the Soviet Union, and Japan—were the most conflictual of the entire two-hundred-year period.[48] The Cold War witnessed significantly more conflict across orders than within them, as did the post–Cold War period. Finally, those pairs of states that were not involved in a trans-order relationship—the "Other" pairs listed in both periods, which did not contain a member of an international order whose core superpower had a veto in the Security Council—exhibited reasonably low rates of conflict initiation by historical standards.

Conclusion

In this chapter, I've argued that international orders are conducive to internal peace but prone to generate external conflict, especially when more than one order exists at a time. I've supported this argument by examining both the historical research that chronicles the waxing and waning of war as well as the cold, hard data that I used in chapter 4 to explore changes in the rate of international conflict initiation.

I also hope to have demonstrated that the mechanisms that I laid out in chapter 7 to explain the complex relationship between international order and international conflict are plausible ones. Although there are extensive literatures on both order and conflict, there is surprisingly little research that connects the two using data. I don't pretend to have had the last word on such an understudied subject, but I do hope to have offered a compelling first cut.

At a minimum, it should be clear that the configuration of international order is a better predictor of conflict in the international system than the spread of peaceful or humanitarian norms. Even if we grant "a big dose of randomness" in the data, a steady *increase* in

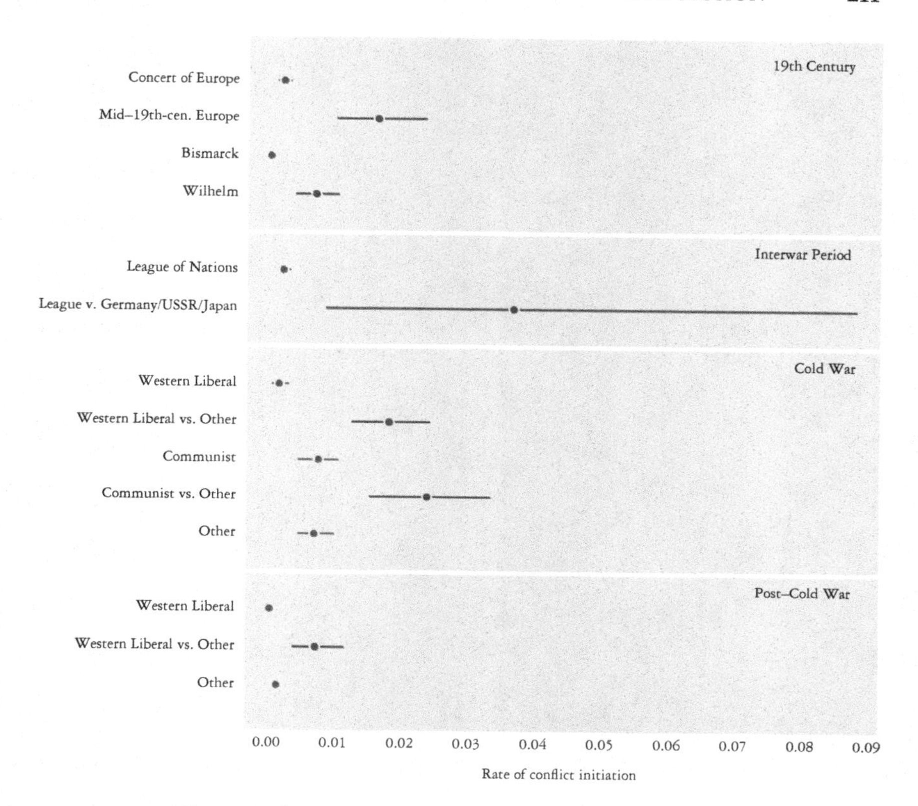

FIGURE 8.1 The conflict-proneness of countries within international orders (blue) relative to those outside of them (red).

the rate of conflict initiation over the course of nearly two centuries is a real problem for the decline-of-war school. By contrast, it lines up very well with the predictions I've made above: The two periods in which a single international order prevailed, the Concert period and the post–Cold War period, are the least conflictual on record, while the post–World War I period and the Cold War, both of which featured multiple international orders, are the most.

PART V
Conclusion

9

Conclusion and Implications

> We are facing increased global disorder, characterized by decline in the long-standing rules-based international order—creating a security environment more complex and volatile than any we have experienced in recent memory. Inter-state strategic competition, not terrorism, is now the primary concern in U.S. national security.
>
> *Summary of the 2018 National Defense Strategy of the United States*

IN THE PREFACE, I told the story of how I got involved in this project: The decline-of-war literature left me with a nagging little itch that something might not be quite right. At the time, I thought it was a simple enough question that I'd get a research note out of it at most. As I told people at the time, "Political scientists try to figure out cause-and-effect relationships for a living. This is just a question of plotting one variable over time and seeing whether it goes up or down. If we can't get *this* right, we really should pack it up and go home."

I've since gained a healthy appreciation for my own naiveté. There were, it turns out, not one but three different ways to look at the question of whether war is in decline. The data demanded sophisticated statistical tests that I either had to brush up on, discover, or create. In the end, no matter how I looked at the data, they simply don't support the decline-of-war thesis. The deadliness of war, whether measured in raw battle deaths or battle deaths normalized by the combatants' population, hasn't decreased in the last two hundred years. A modest decrease in battle deaths as a fraction of world population among smaller wars, most likely due to differential rates of growth between populations and militaries, disappears when wars get larger. There is no systematic upward or downward trend in the potency of the causes of war. And while there is a trend in the rate of international conflict initiation, that trend is toward more rather than less conflict throughout most of the last two centuries. Other than the ends of the two World Wars, only the end of the Cold War produced a decrease in the rate of conflict initiation between countries.

At the same time, it *is* hard to believe that no change whatsoever has happened in the conflict propensity of states. There have been "islands of peace" in the international system over the last two hundred years: the Concert of Europe, the Bismarckian period, the Western liberal order. Those examples suggest a different explanation, focused not on norms or the betterment of humanity but on groups of states that form international orders. A test of that explanation yields strong evidence that order has a salutary effect on the rate of conflict initiation among member states. Unfortunately, the same characteristics that make international orders more peaceful exacerbate their relations with outside states. In the case of single, unchallenged orders, the overall balance of these two tendencies still seems to favor peace in the aggregate. On the other hand, multiple, competing orders—such as those that existed prior to World War II and during the Cold War—are especially prone to conflict.

While this is a very different explanation than the one put forward by the decline-of-war theorists, there are several important points of contact. Although I've been very critical of Professor Pinker's bold claim that empathy and Renaissance humanism are producing a decline in war, I do think that the relationship between Western liberalism and peace is an important part of the story of international conflict in the past century. The post–World War II Western liberal order is the best case for the decline-of-war argument, and it's a very good case. If it made up the entirety of the history of international relations, I would have very little to argue with. (Not nothing—I *am* an academic, after all! But very little.) The problem is that it doesn't. Lots of people even now live outside of the Western liberal order, and it doesn't show obvious signs of expanding to incorporate them any time soon. The willingness, even eagerness, of the West to make war with the East during the Cold War, especially via proxies, dramatically undermines the decline-of-war argument. And as the Concert of Europe and the Bismarckian system demonstrate, illiberal powers can also collaborate to minimize or eliminate war.

Professor Mueller's argument, too, has a lot to recommend it, even though I don't think it has produced the outcomes that he says it has. The peace movements of the 1920s were an incredible thing. We can blame the framers of the League of Nations for an excess of optimism, but that optimism was born in large part of a passionate commitment to the cause of peace and a belief that others shared in that commitment. That said, not everyone believes in *unconditional* peace. Every international order that I've examined here has countenanced war under some set of circumstances. As I write these words, rumors abound that the United States is contemplating what could be staggeringly costly military action to attempt to destroy North Korea's nascent nuclear capabilities, and those rumors are being taken seriously by serious people. At the same time, the UN's R2P doctrine establishes the legitimacy of the use of force in order to prevent human rights abuses. The result in the latter case may be a better world, at least from a Western liberal perspective,[1] but there is no denying that it would be brought about via the threat and use of force.

Professor Goldstein's argument about the impact of peacekeeping on international conflict is both the most specific and in some ways the most challenging argument to assess, largely because the fact that peacekeeping is most likely to happen where it is most needed makes its effects difficult to gauge. It does seem more than plausible, as Professor Fortna (2004*b*,*a*) and others have shown, that peacekeeping significantly decreases the probability that an international conflict will recur. It is also true that, since the end of the Cold War, when the rate of interstate conflict onset did decrease, the efforts of peacekeepers have been directed mostly at subnational rather than international conflicts. Finally, as Professor Fortna (2013) points out, the regions that Goldstein identifies as examples of zones in which conflict has dwindled

(Europe and East Asia, for instance) have seen very little peacekeeping activity.

The argument that ties these threads together—the one ring to rule them all—centers on international order. Liberal humanism and a normative aversion to war have not translated directly into peace, but they have made people more willing to invest in international political orders that moderate the impact of anarchy. The same liberal impulses that lead to peacekeeping have also produced a new justification for war, in the form of humanitarian intervention. And while clashes within international orders are rare, clashes among them are not.

To its great credit, humanity has figured out how to transcend the international war of all against all and create islands of peace. Having done so, though, we don't discard our weapons. We make bigger ones, and we array them along the shore.

Overall Implications

What happens when we pull together the implications of this book for international conflict, international order, and the lethality of war? There are quite a few takeaway points, some more obvious than others.

War Isn't Getting Less Deadly

The most alarming conclusion to come out of this study is that war simply isn't getting less deadly. If they do happen, wars—understood here and elsewhere in the war literature as international conflicts that pass the thousand-battle-death threshold—are no less likely to escalate and kill millions of people than they were in 1913 or 1938. This is true regardless of whether we're talking about deaths in absolute terms or deaths relative to the populations of the combatants. There has been a slight reduction in deaths as a percentage of world population in smaller wars, thanks to the tendency of militaries to grow more slowly than populations, but it hasn't had any impact at all on the propensity of wars to escalate out of control.

To make things worse, when wars snowball, they *really* snowball. Professors Aaron Clauset, Cosma Shalizi, and Mark Newman examine twenty-four phenomena that, according to scholars, get bigger in proportion to how big they already are and in so doing produce power law distributions. The list includes things like frequency of word use in English, severity of terrorist attacks, website hits, wealth, and the sizes of forest fires. For each one the authors estimate the slope coefficient, α, that reflects just how quickly the phenomenon in question snowballs. Only one phenomenon on the list—the intensity of earthquakes—showed a greater tendency to snowball than does war (Clauset, Shalizi, and Newman, 2009, 684).

Very few people who start wars anticipate that millions of people will die in them. These results demonstrate that it is shockingly easy, even now, for millions of people to die in them. When Kaiser Wilhelm II addressed his troops before sending them off to war in August 1914, he famously told them that they would be "home before the leaves fall from the trees." Wilhelm's example, and those of other European leaders who expected a quick and painless war in 1914, would be a good one for any twenty-first-century leader contemplating war to keep in mind: When it comes to the propensity of war to spiral out of control and produce mindboggling death tolls, we live in the same world that they lived in.

International Orders Foster Peace...

... or, if not true peace, at least a very significant reduction in violence. This might seem like an odd statement from a guy who just spent an entire book arguing that war is not coming to an end. But it's true, and it's one of the most important conclusions in the book.

Starting in the post-Napoleonic era, human beings conceived of a form of international order that transcended narrow self-interest. We learned perhaps the most important lesson that humanity can learn: Long-term cooperation is more valuable than short-term gain.[2] The balance-of-power system to which the Great Powers subscribed in the eighteenth century was founded on the idea that a rough equilibrium of power would prevent the worst outcomes that might arise from the unconditional pursuit of short-term national interests. This entailed a modest degree of cooperation—just enough to establish the boundaries within which naked self-interest would be pursued. The Concert of Europe, by contrast, established the principle that the Great Powers should collectively govern the affairs of Europe for the general good and should be willing to sacrifice their own short-term interests in order to do so. This was a revolutionary idea in its own right, and it has persisted in different forms for most of the last two hundred years.

Importantly, there was nothing liberal or progressive about the Concert. Indeed, quite the opposite. Its founders were people who, for the most part, desperately hoped to turn back the clock to a time when liberal ideas of democratic participation in national politics were wholly notional, or at least limited to that weird little island to the north. Nor do it and subsequent orders rely on altruism or empathy. They can be achieved and maintained purely out of rational self-interest, as long as people realize that forgoing some of their short-term goals in order to sustain cooperation is in fact in their interest.

Does the fact that we've figured out how to cooperate and avoid war mean that we can rid ourselves of war entirely if the trend toward international order continues? Unfortunately, the answer to that question is probably "no."

...If We Want It

In late 1971, John Lennon and Yoko Ono bought billboard space in major cities around the United States to put up a simple message in black and white: "War is Over!" Underneath, in smaller letters, was written an important qualifier: "If you want it." They also wove the message into the background of their single "Happy Xmas," released at about the same time.

On one level, the message was an optimistic one. The point that Lennon and Ono were trying to make at the time was that we could end the war in Vietnam by simply not fighting it. But anyone who paid as close attention to Lennon's lyrics as I did when I was growing up had to suspect that there's a darker message as well: *We don't want it.* We could end this war simply by putting our guns down and walking off of the battlefield, but he and everyone else knew that we wouldn't do it. We didn't want the war. But we wanted the alternative even less.

The fact that countries have managed to establish cooperative international orders that reduce violence does not mean that they aren't willing to use force in general. Every international order that I have examined here, and most likely any international order that comes into existence in the future, values *something* more than it values peace. Members of the Concert were willing to wage war to prevent revolution. Communists were willing to fight in order to hasten the demise of capitalism. Western liberal states have been willing to fight to prevent the spread of Communism and to curtail large-scale violations of human rights.

International order does nothing to reduce conflict over issues relating to fundamental values such as these. In fact, as Sir Michael Howard (2002, 2) points out, war may be an intrinsic part of order. That is not inherently a bad thing: the R2P doctrine that obliges the international community to intervene to prevent human rights abuses may very well make the world a better place. But we can't ignore the fact that war and the threat of war are means to that end.

Order Also Produces War

In some ways, international order is actually conducive to more rather than less war. In Tilly's succinct formulation, war made the state and the state made war, and the same is substantially true for international order.

International orders are based on fundamental norms and principles of legitimacy, which in turn inform a collective understanding of what constitute legitimate reasons to go to war. The existence of international order both reinforces and deepens those norms and principles. Moreover, the organization of the states that make up an international order, which helps to alleviate their concerns about one another, also organizes them more effectively to engage in conflict outside the order.

Even when the fundamental principles that underpin order are not at stake, I've shown that international orders produce perverse dynamics that increase the chances of war with countries outside of the order. The existence of order both creates new issues that can result in conflict and alters the calculus of conflict to make war with third parties, and especially with rival international orders, more rather than less likely.

Orders Fade

It's no coincidence, as Professor G. John Ikenberry (2001) and others have pointed out, that international orders tend to form in the wake of major wars. Simply put, there's nothing like millions of deaths to convince people of the value of international cooperation and the importance of pursuing it.

Unfortunately, international orders founded on a particular distribution of power and a common set of norms and principles are typically slow to adjust to changes in those things. The European Concert was founded at the high-water mark of collective conservative sentiment and fear of revolution among the major powers. It functioned most effectively in the few years following its founding. It even managed for some time to survive the substantial ideological rift that emerged in 1830. But as the memory of Napoleon and the urgency of cooperation faded, so too did the Concert.

Today's international order is showing signs of the same sorts of tendencies.[3] The world has changed a lot since 1945, both in terms of the global balance of power and the development of new states and nonstate actors. Along with changes in the balance of power come changes in principles of legitimacy. For all of these reasons, the institutional bargains that were struck at the beginning of the Cold War are becoming increasingly outdated and in need of renegotiation. To take a single emblematic example, reform of the United Nations Security Council to include Germany, Japan, India, and Brazil as permanent members has been debated since the 1990s but has yet to produce concrete results.

At the same time, a recent surge in populist and reactionary sentiment within the core countries of the Western liberal order has, to the surprise of many, put the question of its continued survival on the table. The topic of Canada's spring 2017 Munk debate—a series that brings together prominent scholars and opinion leaders to discuss important issues of the day—was the question of whether the liberal international order had come to an end. At the start of the debate, only 66% of the audience members believed that it had not.[4] The journal *Foreign Affairs* went farther, devoting its March 2017 issue to the question "What Was the Liberal Order?"

While the root causes of the populist and reactionary backlash of the mid to late 2010s are being debated and probably will be for some time, one likely reason for the decline in the appeal of internationalism and cooperation lies in the fact that more than a generation has passed since its benefits were glaringly obvious. Few citizens of Western countries could question the value of the liberal international order while NATO and Warsaw Pact tanks were arrayed against one another in the center of Europe. Fewer still could question the value of a concerted effort toward international order in the wake of World War II. It's not at all clear that international dangers on the horizon—the development of North Korea's nuclear arsenal, Russian revanchism and cyberwarfare, even the rise of China—could represent the sort of existential threat that had previously been the cement holding together the Western order.

That's not to say, of course, that they don't represent potential threats. A conflict in the South China Sea or a crisis with North Korea would be precisely the sort of clash that could escalate dramatically. Partly because these low-probability events *are* far worse than we can imagine, however, we tend not to band together to cooperate until it's too late to prevent them.

That said, if international order is a double-edged sword when it comes to international conflict, there must be pros and cons to the decline of order as well. If the dissolution of the Western liberal order permits conflict among Western developed nations to return, it would also constitute a dissolution of the liberal consensus that underpins humanitarian intervention. As historian Niall Ferguson put it in an interview about the Munk debate, "[t]he thing about nationalists is that they're not particularly interested in getting involved in wars in faraway places, whereas neoconservatives and liberal interventionists in the 1990s and 2000s were all too eager to have boots on the ground" (Slater, 2017).

War Abides

What all of the above adds up to, I think, is that none of the scenarios that realistically faces the world going forward involves the disappearance of war. If, despite the continued existence of a vast network of global political and economic institutions that support and reinforce it, the Western liberal order does disintegrate—or, only slightly more likely I think, devolve to a set of cooperative arrangements across issue areas that deemphasize security—the absence of armed conflict would cease to be taken for granted by a substantial number of countries. That's not to say, of course, that war would immediately break out all over the place. But the liberal international order has done an awfully good job of maintaining peace within its ranks.

If it survives largely unchanged, on the other hand, it seems more than likely that it will continue to use measures up to and including force to reshape the world into something more to its liking. Human rights abuses, terrorism abroad, the spread of weapons of mass destruction, the disintegration of weak states—all of these are causes that have warranted the use of military force quite recently. They're not going away. And the continued existence of the Western liberal order both reinforces the international commitment to fight them and helps to focus and coordinate efforts to do so.

That's not all bad, in and of itself. These are all things that we've decided to fight for, despite the high costs of doing so. Quite a few people believe that making progress against some or all of them *does* make the world a significantly better place. But what we haven't yet done is figure out how to make the world a better place and eliminate war at the same time. As Lennon and Ono pointed out, war *is* over if we want it. We just don't want it. Other things matter more, as they always have.

Unfortunately for us, those things matter to countries with the most powerful military forces and the most lethal weapons in history. Every time one of those countries starts a war, it runs a very small risk that an incomprehensibly vast number of people will be dead by the time the war ends. Over time, the risk accumulates: Every new war is a new roll of the dice. And the odds haven't changed.

APPENDIX

Chapter 4 Notes

What About Civil Wars?

As I mentioned earlier, this book is primarily about international wars, which are typically seen as having different causes than other sorts of wars, especially civil wars.[1] But the decline-of-war literature occasionally discusses trends in civil wars as well. In fact, one of the most widely reprinted graphics presented as evidence of the decline-of-war argument (Figure A1) is a graph that's dominated by trends in civil war.

The graphs in chapter 4 chart the number of armed conflicts per year, of various types: international (or interstate), civil, internationalized civil (civil conflicts in which outside countries become participants), and imperial and colonial. The data are from the UCDP/PRIO project mentioned above. These are the data used by both Pinker (2011*a*, 304) and Mueller (2007, 87; 2014, 46). Pinker includes conflicts with at least 25 battle-related deaths, as do I, while Mueller focuses on conflicts with at least 1,000, but the general trends are remarkably similar.

We've already discussed international conflicts, wars, and disputes as well as extra-state and nonstate wars, and I don't dispute the decline in international conflict around the end of the Cold War, which is really

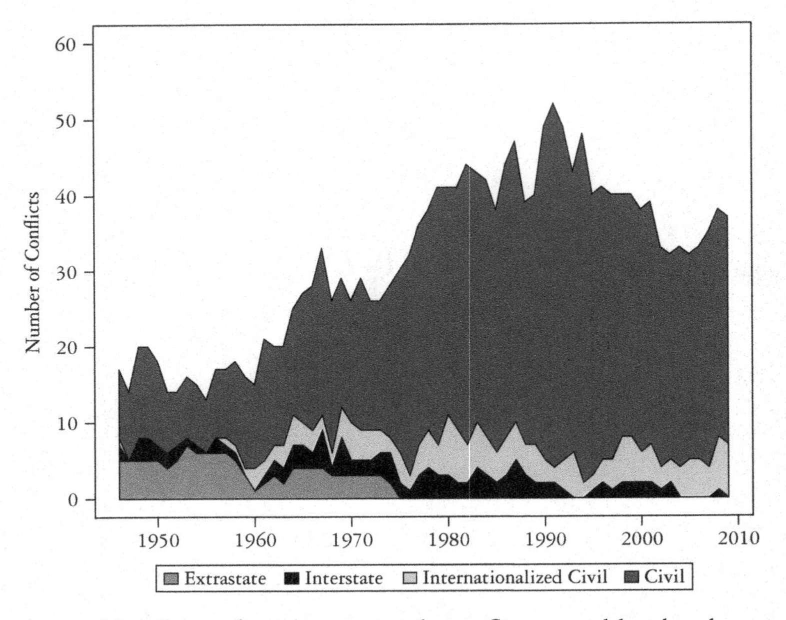

FIGURE A1 Number of violent armed conflicts worldwide, by year. Source: UCDP/PRIO data.

all we can see in this truncated graph. So let's focus on trends in civil war, which seems to plunge after the Cold War as well.

The first thing to note is that this graph measures the number of armed conflicts that are either initiated *or ongoing* in a given year.[2] That's not irrelevant, of course: There is some sense in which the world is more "warlike" if there are many ongoing conflicts than it would be if there were fewer. But the number of ongoing conflicts doesn't really tell us very much about the rate of conflict initiation, and the sources of ongoing conflicts are very different from the sources of conflict initiation.

Take another look at the graph. There are not all that many conflicts in the first few years. That's not surprising: World War II had just come to an end, and systemic wars have a way of resolving old rivalries and setting the stage for new ones. So in the first few years, there are very very few *ongoing* conflicts—most of them are new.

As time passes, however, new conflicts begin to occur while old ones start to pile up. After a few years, ongoing conflicts begin to overshadow new ones, and after a few more years settled conflicts begin to come unsettled. It's *really* not surprising that there aren't many conflicts at the start of the time period, then: Even if the rate of conflict initiation remained constant or declined slightly, we'd expect to see ongoing conflicts start to pile up over time.

I should note that this is actually *good* news for the decline-of-war thesis: The ramp-up in civil conflict-years during the Cold War is really something of an artifact! It doesn't reflect an increase in conflict initiation, any more than a gradually filling bucket reflects the rate at which water is poured into it. In fact, if we give the bucket a slow leak to represent the rate at which conflicts are resolved, we have a pretty good metaphor for how the data in these graphs were generated, and the fact that the bucket is getting full doesn't tell us anything about changes in the rate at which water is being poured into it.

We can also be pretty sure that the rate at which conflicts were resolved increased substantially after the Cold War, which makes sense: During the tightly polarized Cold War, quite a few proxy conflicts were sustained by dollars and weapons from the two superpowers. Once the Soviet Union dissolved, the balance shifted in these proxy conflicts, and the weaker side had every reason to capitulate. In other words, the hole in the bottom of the bucket abruptly got a lot bigger, and most of the water it had been holding rushed out over the next few years.

Unfortunately, though, what that means is that the abrupt decline in conflict-years following the end of the Cold War also doesn't tell us much about the rate of conflict initiation. We can tell that, for a time, more water flowed out of the bucket than into it, but we can't tell whether the amount flowing in dwindled or just failed to keep pace.

In order to get a sense of the answer to that question I consulted a former graduate student of mine, Professor Benjamin Jones, whose dissertation on civil wars had used the same dataset as the books by Pinker and Mueller. As part of his research on the dynamics of civil war initiation and termination, Jones distinguished between new and ongoing fighting in both new and recurring civil conflicts since 1950.[3] I've graphed the result in Figure A2.

This perspective lets us get a much better handle on the frequency of conflict initiation (the dark blue and dark red bands) throughout the

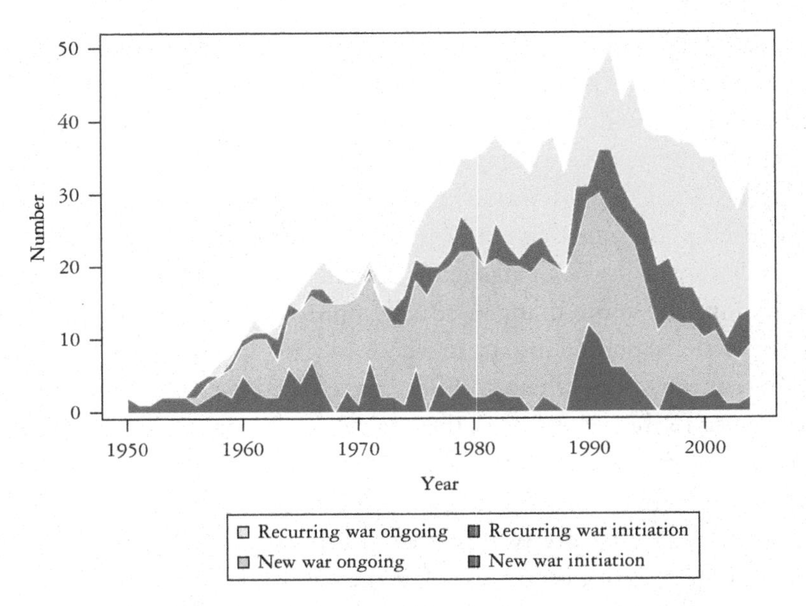

FIGURE A2 Number of civil wars worldwide, by year and type.

Cold War and early post–Cold War period. As the figure demonstrates and intuition suggests, the considerable majority of the conflict years starting in about 1960 reflect ongoing conflicts rather than new ones.

There is one more objection to take into account: Like Figure 2.1, none of these figures actually represent the rate of civil war initiation because none take into account the number of conflict opportunities. The easiest way to correct that problem is to assume that the number of possible civil wars in the international system at any given point in time is directly proportional to the number of countries in the system at that time, and simply divide the number of initiations by the number of countries. It may seem a bit odd to include countries like the United States and Canada in the denominator, since civil war seems unlikely to break out in either one. But the fact that civil war doesn't break out in either of them during this period is relevant to the worldwide rate of civil war initiation over time.

It's important to emphasize that dividing by the number of countries in the international system will not disadvantage the decline-of-war hypothesis—in fact, quite the opposite. The number of countries in the international system more than doubles over this time period, from

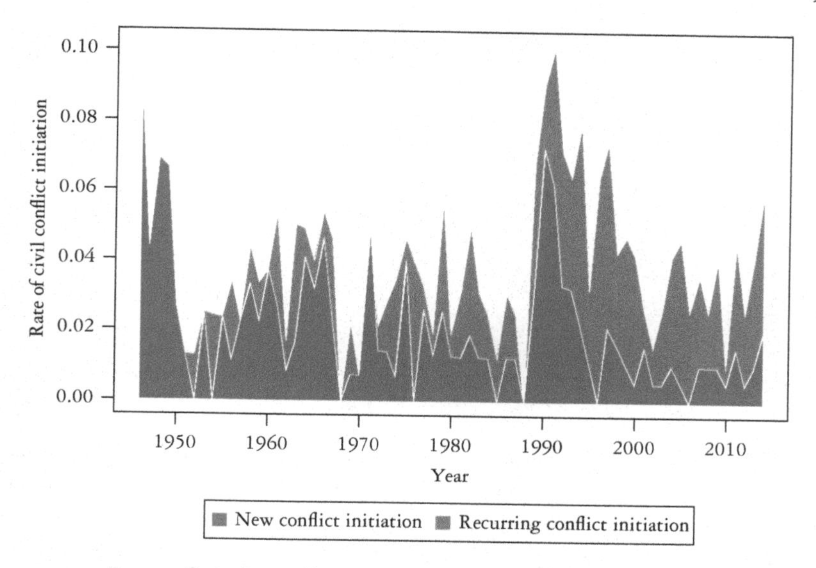

FIGURE A3 Rate of civil conflict initiation worldwide, by year and type.

seventy-five countries in 1950 to 192 countries in 2004. Because the number of countries is in the denominator, that means that the rate of conflict initiation should show more of a downward trend than the raw frequency of conflict initiation (which is what Mueller and Pinker use) would have.

What happens when we divide the number of civil war initiations per year by the number of countries in the international system in that year in order to get a better look at the rate of conflict initiation? To answer that question, I went back to an updated version of the dataset used in the above figures—the UCDP/PRIO Armed Conflict Dataset.[4] I differentiated between new and recurring conflict initiations, as above, but omitted ongoing conflict years. The result is Figure A3.[5]

Unsurprisingly, we see that as time goes on, the balance shifts away from the initiation of new civil conflicts and toward the initiation of recurring ones. We also see, at least in the early years of the post—Cold War period, a surge in the number of recurring conflict initiations—a fact that, on the surface at least, could represent a problem for Goldstein's argument that better peacekeeping has reduced the number of conflicts following the Cold War.

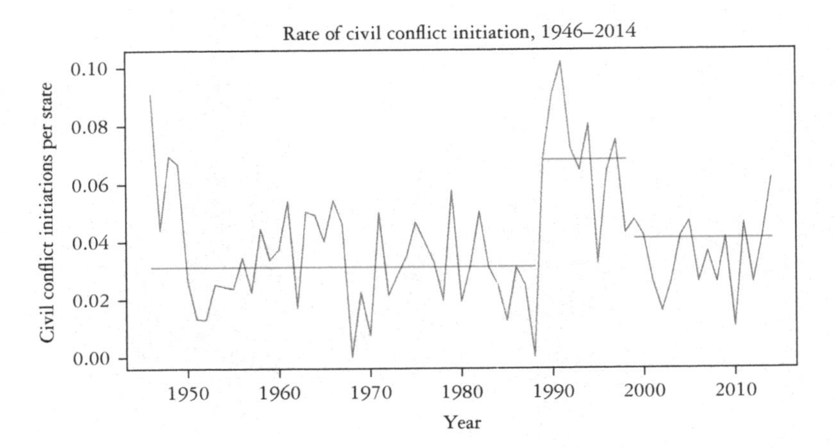

FIGURE A4 Rate of civil conflict initiation worldwide, with change-point test.

In fact, even the most diehard optimist would find it difficult to construe Figure A3 as good news for the decline-of-war thesis. For the most part, it's hard to see a pattern at all during the Cold War, and to the extent that one can be discerned we actually see an *increase* rather than a decrease in the rate of civil conflict initiation after the fall of the Iron Curtain. Professor Stathis Kalyvas and Dr. Laia Balcells (2010) noted this trend and attribute it to the end of the Cold War, which resulted in a dramatic decline in funding for rebels around the world, an increased reluctance on the part of the United States to prop up weak clients, and the dissolution of the Soviet Union and Yugoslavia (see also Cederman, Buhaug, and Rod, 2009; Howard and Stark, 2018).

Figure A4, which shows the results of applying the change-point detection algorithm to the civil conflict initiation data from Figure A3, confirms that impression. The rate of civil war initiation roughly doubled at the end of the Cold War, though it dropped significantly about a decade later. That decrease, though, looks much more like a return to the previous rate of initiations than any kind of evidence of a pervasive and enduring peace.

There's one caveat to this conclusion, however, and it's a big one. The original UCDP project started gathering these data in 1989, and the 1945–1988 data were only gathered later, from secondary sources. As

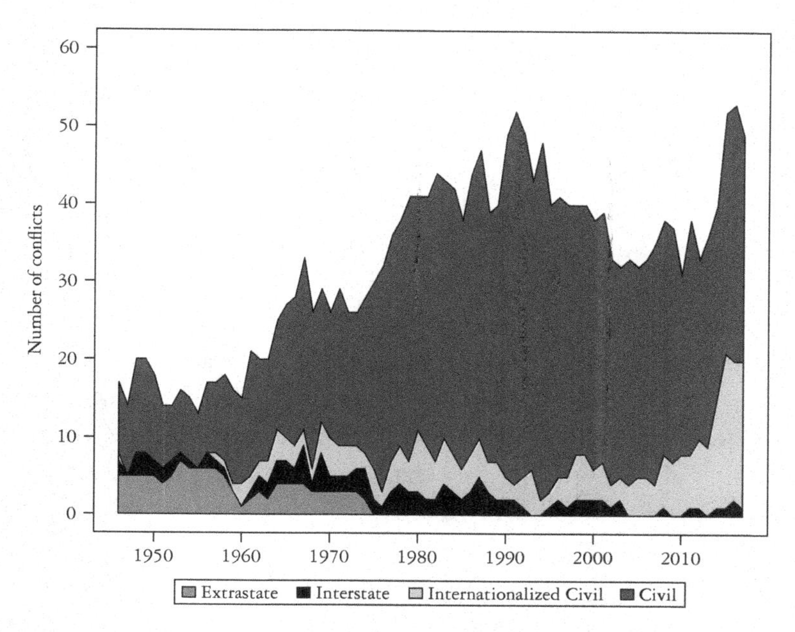

FIGURE A5 Number of state-based armed conflicts by type, 1946–2017.

a result, the material used to code conflict episodes prior to 1989 was very different from the material used to code such episodes after 1989. Because the cutoff between the two coincides perfectly with the end of the Cold War, it's difficult, perhaps even impossible, to ascertain how much of the increase in conflict in the data is real and how much is just an artifact of differences in coding.[6] The most prudent conclusion, I think, is that we simply don't know the answer one way or the other.

Finally, even if you don't believe anything *at all* that I've written in the last few pages about initiations and ongoing wars and are convinced that counting up the number of active conflicts in each year is the right way to determine whether or not conflict is in decline, there's a much more practical reason not to believe the decline-of-war theorists: As the most recent annual tally from the *Journal of Peace Research* (Pettersson and Eck, 2018) demonstrates, the claim hasn't held up well at all in recent years. In fact, if you're inclined to believe this measure, you'd be forced to conclude that 2016 was the most bellicose year on record since the end of World War II (Figure A5).

What About Political Relevance?

Another obvious question is whether the measure of political relevance that I use, which I also helped to create (Braumoeller and Carson, 2011), isn't perhaps skewing the results in some way. To explore that possibility, I have re-calculated these trends using the industry-standard measure of political relevance, derived from Maoz and Russett (1993): all contiguous pairs of countries or pairs of countries within 150 miles of one another by water, and all pairs of countries that include major powers, are defined as politically relevant while all pairs that meet neither criterion are defined as politically irrelevant. As we can see in the left-hand column of Figure A6, this change makes virtually no difference whatsoever. The only change occurs in the rate of the use of force in the 1850s–1860s, when a brief spike in conflict raises the median somewhat before it returns to a lower level. There is still a drop in the rate of the use of force following the Cold War, but the Cold War is still the most conflictual peacetime period on record, and the overall trend throughout the period is still toward more rather than less frequent uses of force.

I also constructed a hybrid measure of political relevance, one in which minor-minor pairs were assigned the Braumoeller-Carson measure of political relevance but any pair of countries containing a major power was assumed to be relevant. This is the "modified Braumoeller-Carson" metric in the right-hand column. This measure is meant to address the not-unreasonable criticism that (a) pairs of countries that include a major power should automatically be considered potentially relevant, no matter what the model says, and (b) there are a lot of them in the international system. This modification highlights exactly the same trends as the Maoz-Russett measure.

The trends in the initiation of all-out war (bottom row) show no meaningful change at all when the different measures of political relevance are applied.

What About the Correlates of War Data?

Another worthwhile robustness check has to do with the Correlates of War data themselves. The Correlates of War project is a well-funded, multi-decade endeavor, and its data on militarized disputes and war

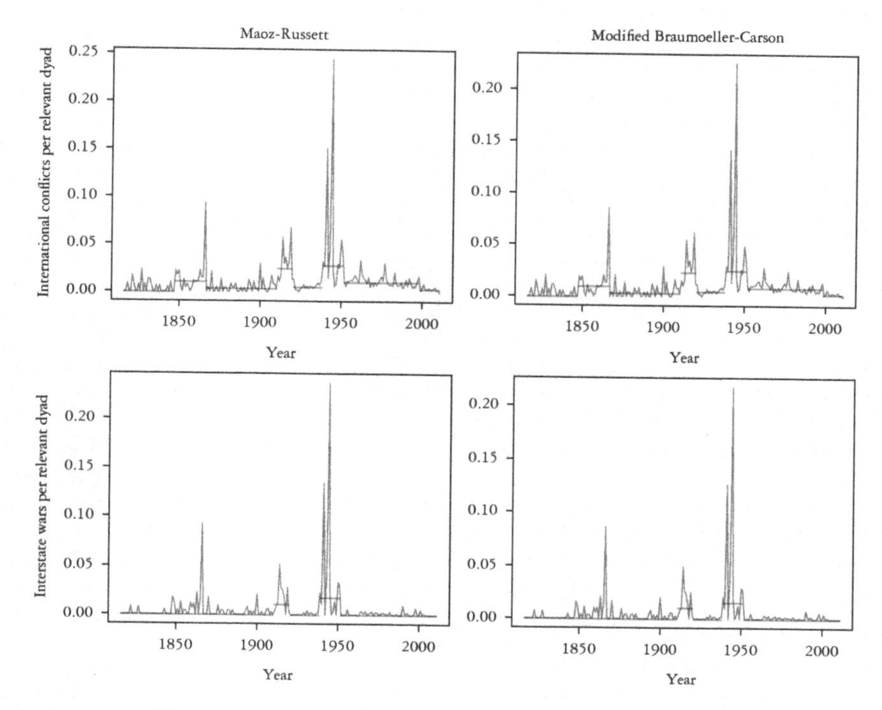

FIGURE A6 Change-point analysis of Correlates of War use of force (top row) and interstate war (bottom row) data, 1816–2014, with different measures of political relevance. The light gray trend line charts the rate at which interstate war occurs; the horizontal red lines indicate medians. Breaks between the horizontal lines indicate points at which the change in the overall trend is larger than one could reasonably expect by chance.

form the backbone of international conflict studies in political science. As I was in the process of writing this book, however, I became aware of an alternative war dataset covering the same period, the Interstate War Dataset (IWD), created by Professors Dan Reiter, Allan Stam, and Michael Horowitz (Reiter, Stam, and Horowitz, 2016). As the authors note in their online Appendix, "Though this data set builds on the Correlates of War (COW) 4.0 interstate war data, it is not in any way associated with or sanctioned by COW."[7] Despite nontrivial differences between the two datasets,[8] the Reiter/Stam/Horowitz IWD data show the same aggregate pattern of war initiation over time (Figure A7). The latter half of the nineteenth century does look a bit more conflictual

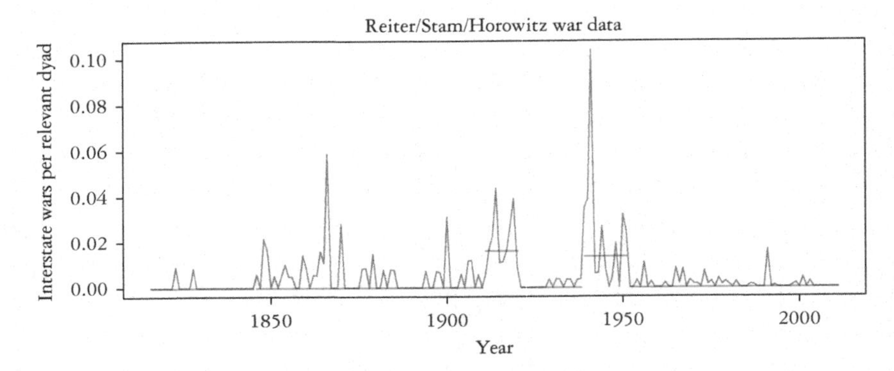

FIGURE A7 Change point analysis of Reiter/Stam/Horowitz interstate war data, 1816–2007. The light gray trend line charts the rate at which interstate war occurs; the horizontal red lines indicate medians. Breaks between the horizontal lines indicate points at which the change in the overall trend is larger than one could reasonably expect by chance.

than it does in the COW data, but the order of the non-World War periods from highest to lowest rate of initiation—second half of the nineteenth century, Cold War, interwar period, and first half of the nineteenth century—remains the same.[9]

What About a Poisson Test?

Professor Pinker, following in the footsteps of Professor Lewis Fry Richardson (1960*b*), argues that the outbreak of international conflict follows what's called a *Poisson distribution,* a classic description of the expected number of events per unit of time given that those events occur with a certain average rate but otherwise occur entirely at random. The canonical application (Von Bortkewitsch, 1898) was to the number of deaths by horse or mule kicks in fourteen corps of the Prussian Army. Professor Richardson was the first to argue that the outbreak of international conflict, too, follows a Poisson distribution, prompting Professor Pinker (2011*a*, 207) to observe that "[t]he possibility that war might decline over some historical period, then, is alive. It would reside in a nonstationary Poisson process with a declining rate parameter." This would seem like a fairly obvious place to start, but I haven't mentioned it at all. Why not?

The short answer is that statistics has come a long way since 1960 (or 1898). It'd be possible to look for the evidence of decline in this manner, by running a Poisson regression with time as an independent variable and the number of conflict opportunities (technically, the log of the number of conflict opportunities) as the offset parameter. The problem is that you're making two big assumptions—that the data follow a Poisson distribution, and that the decline over time is linear—that are probably wrong. Unfortunately, because those assumptions are "baked in" to the statistical technique, if they're wrong, then the statistical results will probably be wrong, too.

Fortunately, it's possible to relax some of these assumptions. One of the most restrictive is the assumption, built into the Poisson distribution, that the mean of the distribution of the data is equal to the variance. We can estimate the relationship between the mean and the variance of the data rather than assuming it by using the more flexible and general quasi-Poisson distribution. We can also add polynomial terms ($year^2$, $year^3$, etc.) to allow for nonlinear change in the baseline rate over time.

Doing these things produces results that are only moderately interesting. A simple quasi-Poisson regression of rate on time fails to support the claim that the rate of conflict initiation has declined over time.[10] If we relax the linearity assumption, we find a significant, curvilinear relationship between time and rate of conflict onset: that rate increases steadily until the early twentieth century, then declines thereafter. All that result tells us, really, is that there were two World Wars in the middle of the period, during which the rate of conflict initiation skyrocketed.[11] It doesn't suggest more nuanced results because a model this simple *can't* suggest more nuanced results.

Anyway, here's my point: If you continue down this road of relaxing statistical assumptions and allowing for a more nuanced depiction of the relationship between time and the rate of conflict onset, *you end up with exactly the test that I use in Chapter 4*. The change-point detection algorithm that I used in that chapter (James and Matteson, 2014) does not assume that the data follow a Poisson distribution—or any distribution at all. It does not assume that the trend over time is linear or curvilinear; it can be just about any trend at all. Relaxing these assumptions requires a lot of computing power, which Richardson didn't have at his disposal. I do.

So while it may appear that I've ignored the suggestion of a Poisson test, I've actually done anything but. I started with a test along those lines, but I soon realized that its assumptions were too restrictive to give useful results and opted for a more modern and more powerful statistical test instead. That test does *exactly* the same thing that the Poisson test that Professor Pinker proposed is intended to do: look for changes in the systemic rate of conflict initiation over time. It just does a better job.

Chapter 5 Notes

Regression Results

I know, I know, I said I'd avoid math. But the discussion of the relationship between population, military size, and war deaths from chapter 5 relied mainly on a scatterplot that was just *begging* for a regression analysis. To avoid answering the same question from my colleagues over and over again, here it is.

The table presents two models—one that directly compares the impact of time and population on war intensity, and a second that breaks up time into three periods that match the periodization in Figure 5.11. In both, the dependent variable and the first independent variable are just what they say they are: battle deaths divided by the pooled populations of the combatants, and the pooled populations of the combatants, respectively. If Oka et al. (2017) are correct that larger societies have smaller militaries and therefore fewer conflict fatalities relative to their size, the coefficient on this first variable should be negative—and it is. It's also strongly statistically significant, indicating that it's very unlikely to have occurred by chance.

The second variable in the first model just measures the year, and the second and third variables in the second model are dummy variables for the 1901–1945 and post-1945 periods, respectively. The omitted category is 1815–1900. If the relationship between state size and military fatalities is an artifact of time—that is, if what's really happening is that countries are getting both bigger and more peaceful over time—we should expect to see negative and statistically significant coefficients for these variables as well. The overall relationship, captured by the year variable, is in fact

TABLE A.1 Regression Results

	Dependent Variable	
	log(War Intensity)	
	(1)	(2)
Population	−0.563***	−0.554***
	(0.129)	(0.123)
Year	−0.004	
	(0.004)	
1901–1945		0.983
		(0.518)
Post-1945		−0.487
		(0.489)
Constant	11.897	4.288**
	(8.223)	(1.354)
Observations	95	95
R^2	0.212	0.277
Adjusted R^2	0.195	0.253
Residual Std. Error	2.003 (df = 92)	1.930 (df = 91)
F Statistic	12.413*** (df = 2; 92)	11.597*** (df = 3; 91)

Note: *p<0.05; **p<0.01; ***p<0.001

negative, but the coefficient is about the same size as the standard error, indicating that the change over time in battle deaths as a percentage of pooled population can't be distinguished from chance variation once the size of the pooled population has been taken into account.

This regression is for descriptive purposes only. As I argue extensively in chapter 5, the fact that the dependent variable follows a power-law distribution means that it requires special treatment, and logging it is a very imperfect way of doing that. That said, what it shows is precisely what Oka and his colleagues argued: Larger countries fight smaller wars, relative to their population size. It also shows that the passage of time has no effect on the size of wars when population size has been taken into account.

A Formal Test for Differences in the Slopes of Power-Law Distributions

Statistical tests relating to power-law distributions are surprisingly recent developments. For most of the history of research on power-law-distributed phenomena, it seems, researchers have been content to find a roughly linear pattern in a log-log plot and call it a day. A very smart paper by Professor Aaron Clauset and his colleagues (Clauset, Shalizi, and Newman, 2009) is the first, to my knowledge, to present a comprehensive test of whether a phenomenon conforms reasonably well to a power-law distribution *and* whether another thick-tailed distribution, such as the exponential or log-normal, provides a better fit.[12]

It is perhaps no surprise that, when I began work on this section, I wasn't able to find a useful test to assess whether there was a statistically significant difference between the slopes of two power-law-distributed sets of data. The Clauset et al. procedure was quite recent, and most of the attention in the complex-systems literature seems to be on assessing fit and measuring the slope.

Because I wasn't able to locate a procedure, I came up with one myself. It's very simple, and the simulations below show that it works very well. The purpose of this section is to explain, for the serious statistics nerds, how the procedure works.[13]

The Clauset et al. procedure involves three steps: (1) estimate the lower bound of the power-law relationship, x_{min}, such that all data $\geq x_{min}$ give the best possible fit to a power-law distribution; (2) estimate the slope coefficient, α, for all data $\geq x_{min}$ using maximum likelihood; (3) resample the data with replacement many times and repeat steps 1–2 in order to bootstrap the empirical sampling distribution of the slope coefficient, from which we can derive confidence intervals for the slope coefficient.

Testing for a change in the coefficient at time t is not discussed in the paper, but extending their methodology to accomplish this goal isn't difficult. In a nutshell, I implement the basic procedure on two subsamples of the data—the data at time $< t$ and the data at time $\geq t$—and calculate the difference in coefficients across the two subsamples. That difference can then be bootstrapped to provide an empirical sampling distribution. The hypothesis of a change in coefficients amounts to a claim that zero lies outside the 95% confidence intervals of the empirical sampling distribution of the difference in slope coefficients.

For this technique to work, it must have good coverage, meaning that the true value of the test statistic (i.e., either the slope coefficient or the difference in slope coefficients, depending on which test we're running) has to be within the estimated 95% confidence intervals roughly 95% of the time in repeated samples. In order to assess the technique therefore, I used the code below to generate many sets of random power-law-distributed data with known coefficients, estimated the test statistic and bootstrapped the confidence intervals for each dataset, and checked to ensure that the true coefficient lay within the 95% confidence intervals about 95% of the time.

R does not have a built-in power-law distribution function, so before doing anything else, I load the existing poweRlaw library, which has the rpldis() command for generating random power-law-distributed data.

```
library(poweRlaw)
```

Next, I create a function, calc.pl.coefs(), to estimate x_{min} and α. It estimates x_{min} by iterating across all possible values of x_{min}, estimating α for all values $\geq x_{min}$ at each iteration, and using the Kolmogorov-Smirnov test to assess the fit between the data and a theoretical power-law distribution truncated at x_{min} with a slope coefficient of α. The function takes the data as an argument and returns estimates of x_{min} and α.

```
calc.pl.coefs <- function(mydata){
      xmins <- unique(mydata) # get all unique values of
          data
      dat <- numeric(length(xmins))
      z <- sort(mydata)
      for (i in 1:length(xmins)){
            xmin <- xmins[i] # choose next xmin
                candidate
            z1 <- z[z>=xmin] # truncate data below this
                xmin value
            n <- length(z1)
            a <- 1+ n*(sum(log(z1/xmin)))^-1 # estimate
                alpha using direct MLE
            cx <- (n:1)/n # construct the empirical CDF
            cf <- (z1/xmin)^(-a+1) # construct the
                fitted theoretical CDF
            dat[i] <- max(abs(cf-cx)) # compute the KS
                statistic
```

```
        }
        D <- min(dat[dat>0],na.rm=TRUE) # find smallest D
            value
        xmin <- min(xmins[which(dat==D)]) # find xmin value;
             use min in case of tie
        z <- mydata[mydata>=xmin] # truncate data below xmin
        z <- sort(z)
        n <- length(z) # get tail length (in case we want to
             know that)
        alpha <- 1 + n*(sum(log(z/xmin)))^-1 # get
            corresponding alpha estimate via MLE
        return(c(xmin, alpha))
}
```

To make sure that we're in the ballpark, we can generate some random power-law-distributed data with known parameters and then use the function to estimate the parameters.

```
fakedata <- rpldis(250, xmin=10, alpha=1.8) # Generate data
calc.pl.coefs(fakedata) # Estimate coefficients
```

The resulting estimates on a representative run were $x_{min} = 12$ and $\alpha = 1.79$, indicating that the routine does a good job of recovering the coefficients. (There will always be a little error, and since randomization is involved, your results will differ from mine, but they should be very close.)

To generate the empirical sampling distribution, we bootstrap it from the data.

```
bootstraps <- NULL
for(b in 1:500){
        replicate <- sample(fakedata, size=length(fakedata),
             replace=TRUE)
        bootstraps <- c(bootstraps, calc.pl.coefs(replicate)
            [2])
        }
hist(bootstraps)
```

Then we can generate percentile confidence intervals from the empirical sampling distribution.[14]

```
c(quantile(bootstraps, 0.025), quantile(bootstraps, 0.975))
```

In my sample run, these come out to $(1.69, 2.10)$, though again, your mileage may vary.

So that's the basic technique.

Next, in order to check coverage, I generate a thousand simulated datasets. For each dataset I use a thousand bootstrap replicates to generate an empirical sampling distribution and calculate 95% c.i.s. Finally, for each dataset I record whether or not the 95% c.i.s contain the true parameter. If the estimator is working correctly, the c.i.s should contain the true parameter about 95% of the time.

Note that, unsurprisingly, this takes awhile to run. It would be straightforward to parallelize the code using the boot() command from the bootstrap library, but I wanted to make this code as transparent as possible.

```
cover <- NULL # Set up vector to hold coverage (true/false)
     for each individual test

for(iter in 1:1000){ # Do the following 1,000 times:
       fakedata <- rpldis(250, xmin=10, alpha=1.8) #
           Generate random PL-distributed data
       bootstraps <- NULL
       for(b in 1:1000){ # Estimate alpha in 1,000
           resamples to get empirical sampling distribution
              replicate <- sample(fakedata, size=length(
                  fakedata), replace=TRUE)
              bootstraps <- c(bootstraps, calc.pl.coefs(
                  replicate)[2])
       }
       cover <- c(cover, (quantile(bootstraps, 0.025) < 1.8
            & 1.8 < quantile(bootstraps, 0.975)))
} # Register TRUE if true parameter is between the 95% c.i.
    s; register FALSE otherwise.

(sum(cover)/length(cover) * 100) # In what percentage of
    cases did the 95% c.i.s contain the true parameter?
```

My representative run found that the 95% confidence intervals contained the true parameter 94.9% of the time. The coverage of confidence intervals from the nonparametric bootstrap actually shouldn't be *quite* that good, simply because bootstrap estimates of confidence intervals are approximate. So I lucked out on this particular run.

Now I run the same test for the difference between two coefficients. First, I run a placebo test: all of the data come from the same data-generating process. The true difference between slope coefficients, 0, should be within the 95% c.i.s 95% of the time. Again, this takes awhile to run.

```
cover <- NULL # Set up vector to hold coverage values

for(iter in 1:1000){ # Procedure is the same, except that
    the test statistic is the difference between the slopes
     in two subsets of the data.
       fakedata <- rpldis(500, xmin=10, alpha=1.8)
       bootstraps <- NULL
       for(b in 1:1000){
```

```
            replicate1 <- sample(fakedata[1:250], size=
                length(fakedata[1:250]), replace=TRUE)
            replicate2 <- sample(fakedata[251:500], size
                =length(fakedata[251:500]), replace=TRUE
                )
            bootstraps <- c(bootstraps, (calc.pl.coefs(
                replicate1)[2] - calc.pl.coefs(
                replicate2)[2]))
        }
        cover <- c(cover, (quantile(bootstraps, 0.025) < 0 &
             0 < quantile(bootstraps, 0.975)))
}

(sum(cover)/length(cover) * 100) # Range should cover true
    alpha 95% of the time.
```

Here, the sample run has a coverage rate of 97.8%—a bit on the high side, but not too surprising since the difference between two approximations will be even more approximate.

Finally, I run the same test for the difference between two coefficients that differ. The true difference, 0.4, should be within the 95% c.i.s about 95% of the time.

```
cover <- NULL # Set up vector to hold coverage values

for(iter in 1:1000){
        fakedata.1 <- rpldis(250, xmin=10, alpha=2.2)
        fakedata.2 <- rpldis(250, xmin=10, alpha=1.8)
        bootstraps <- NULL
        for(b in 1:1000){
            replicate1 <- sample(fakedata.1, size=length
                (fakedata.1), replace=TRUE)
            replicate2 <- sample(fakedata.2, size=length
                (fakedata.2), replace=TRUE)
            bootstraps <- c(bootstraps, (calc.pl.coefs(
                replicate1)[2] - calc.pl.coefs(
                replicate2)[2]))
        }
        cover <- c(cover, (quantile(bootstraps, 0.025) < 0.4
             & 0.4 < quantile(bootstraps, 0.975)))
}

(sum(cover)/length(cover) * 100) # Range should cover true
    alpha 95% of the time.
```

Here we get a coverage rate of 97.6%—again, slightly high, but not horrendously so. High coverage means that, at the margin, we're *slightly* more likely than we should be to fail to reject the null hypothesis of no difference—so if I were to encounter *p*-values of around 0.06 or 0.07, I'd be inclined to try to come up with a more precise test. In the event, none of the *p*-values from any of the tests reported above were even close to statistical significance, so the question was moot.

Chapter 6 Notes

Detailed Results

While Figure 6.1 gives us a sense of aggregate trends within and across categories, it is somewhat lacking in specific detail. The plots in this Appendix present that detail for those who are interested. Following convention, the gray areas surrounding the main trend lines represent 95% confidence intervals, which capture our degree of uncertainty about the estimates.

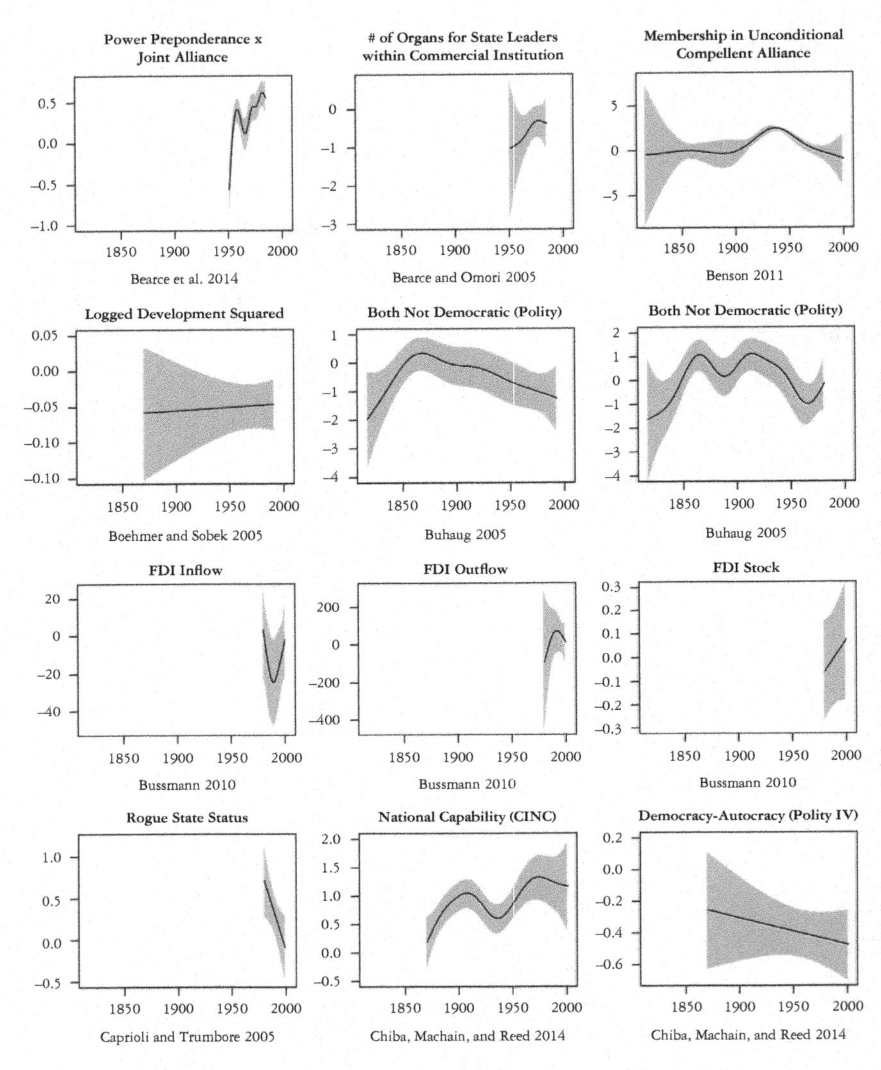

FIGURE A8 Changes in the potency of the causes of war, 1815–present (detail).

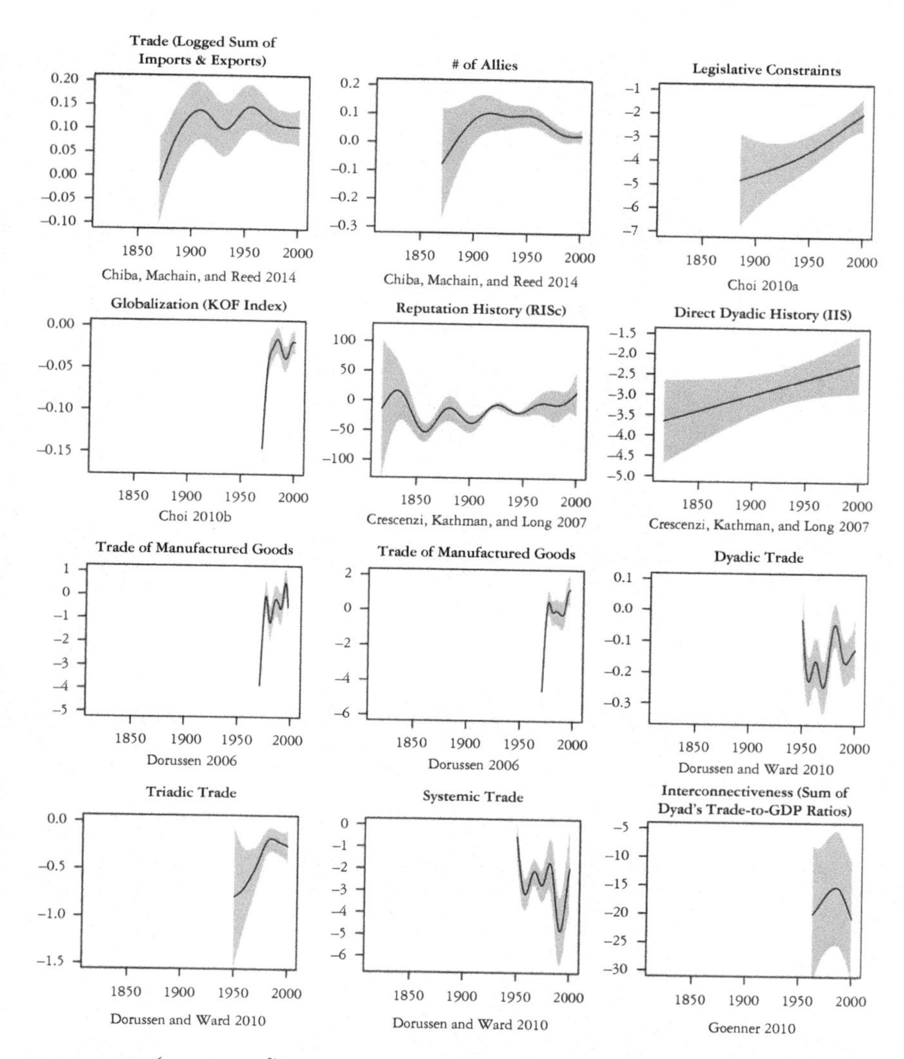

FIGURE A8 *(continued)*

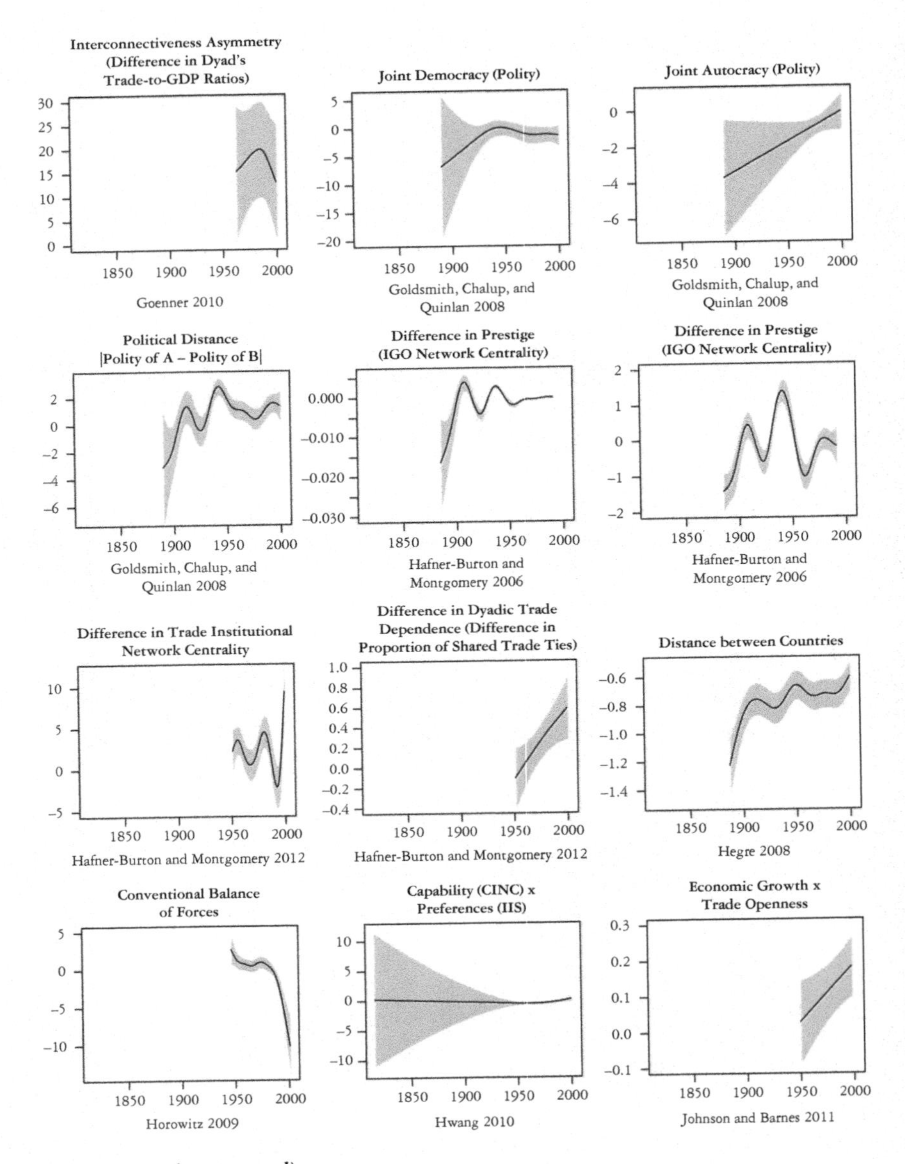

FIGURE A8 *(continued)*

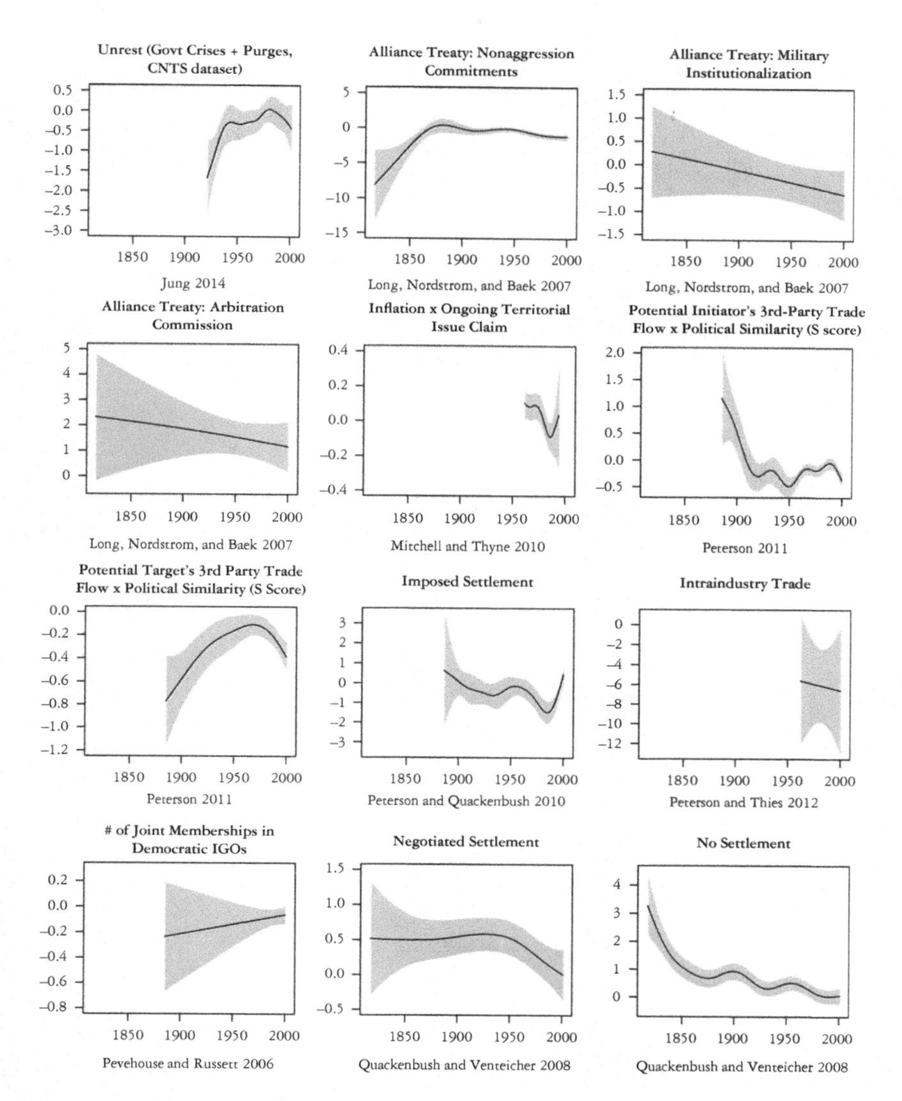

FIGURE A8 *(continued)*

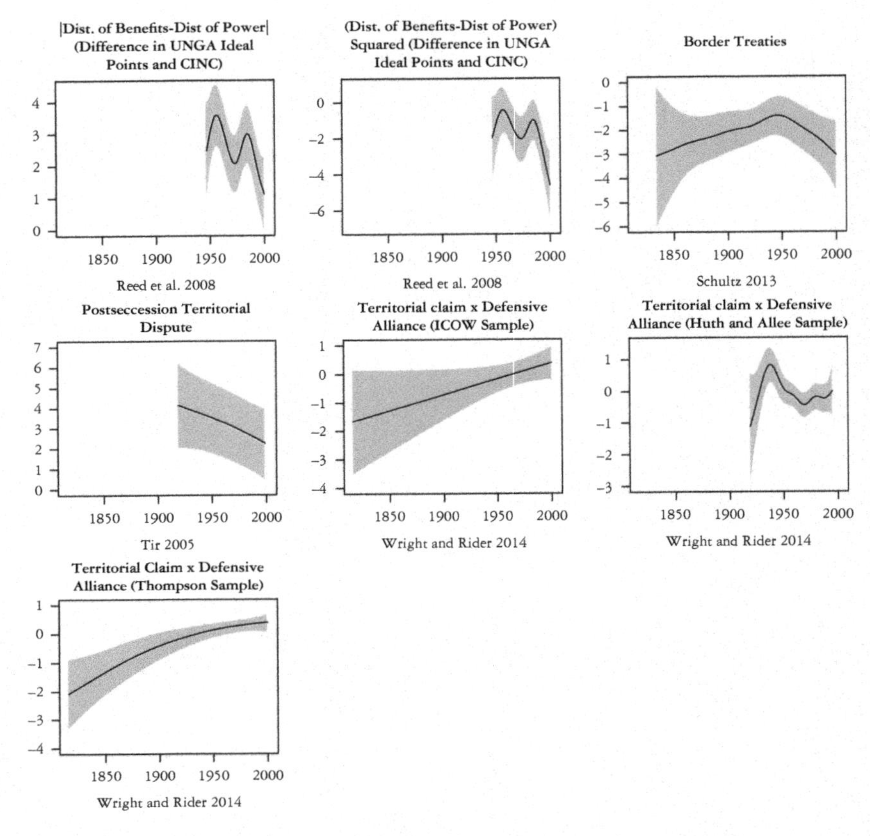

FIGURE A8 *(continued)*

NOTES

Preface

1. n countries have $\frac{n(n-1)}{2}$ potential connections with one another. So if there are 195 member countries in the United Nations, there are almost nineteen thousand different ways they could pair up.
2. See, e.g., Fortna (2013); Levy and Thompson (2013); Levy (2013); Mearsheimer and Walt (2013); Gray (2015); Cirillo and Taleb (2016*a*); Clauset (2018).

Chapter 1

1. Fatality data are from White (2013), whose sources are listed exhaustively at his website http://necrometrics.com, and from Goldstein (2011, 263—264). Any discussion of data on war deaths must include a hefty caveat: Estimates of fatalities due to war are notoriously difficult to calculate and can be quite controversial. For example, while the Wikipedia page for the Second Congo War lists a total of 5.4 million fatalities, Goldstein (2011, 260–263) offers a detailed rebuttal. Given the uncertainty of the estimates, however, Goldstein admits that "the real numbers could be way above or below the estimates of 2.8 million or 0.9 million."
2. http://democraticpeace.wordpress.com/2008/11/24/reevaluating-chinas-democide-to-73000000/
3. United Nations High Commissioner for Refugees (2017).
4. Especially, but not exclusively, Mueller (1989, 2007); Pinker (2011*a*, 2018); Goldstein (2011).

5. Measuring the deadliness of war relative to world population *does* produce a decline in deadliness over time, but as I argue in chapter 5, that's probably just due to the fact that militaries grow more slowly than populations, and it has no effect at all on the propensity of small wars to snowball into enormous ones.
6. The quote is from page 102 of Santayana's *Soliloquies in England and Later Soliloquies* (1922). Although it is often attributed to Plato, it wasn't by Santayana, who to the best of my knowledge is the originator of the quote.
7. It's worth noting that, if we accept the premise that nuclear weapons promote peace, the normative prohibition against war could also be said to be essentially irrelevant.
8. Richardson (1960*b*). Polish statistician Ladislas Von Bortkewitsch (1898) demonstrated that many quantities that fit this description—the number of deaths by suicide of Prussian children under the age of ten, the number of fatal accidents in labor associations, and most memorably the number of deaths by horse-kick in Prussian army corps—follow a Poisson distribution.
9. Albertini (1980), Dedijer (1966), and Taylor (1954) are classic accounts of the events leading up to the war. For an outstanding account based on more recent scholarship see Clark (2013).
10. Accounts of the assassination vary, sometimes widely, in more specific details—which assassins had which weapons, whether the cars began to back up or had to be pushed due to the absence of a reverse gear, and so on—but they are fairly consistent on these points. I have relied primarily on the account of Clark (2013, 370–376), which subsumes and expands upon earlier versions.
11. This one is more obscure than most; see p. 140 of Brecher and Wilkenfeld's monumental *A Study of Crisis* (1997) for details.
12. Halberstam (2001, 606). I am grateful to Michael Lopate for connecting the spark–powder keg analogy to Bundy's quote.
13. See Pinker (2015*a*) and Goldstein and Pinker (2016) for examples of such conclusions.
14. I should note that this is very much a political scientist's view of the origins of the state, dating back at least to Thomas Hobbes's *Leviathan* (1651).

Chapter 2

1. It's possible to argue, of course, that conflict opportunities are always present regardless of where national borders are drawn. Two feuding polities could still fight once they're brought within the confines of a larger state or empire, so the fact that they don't do so is evidence of pacification. In fact, Pinker makes this argument on pp. 235–236 when he points out that, as smaller polities were absorbed into larger kingdoms, their wars with one another were quelled by the pacifying effect of the civilization process. As a result, he argues, the decrease in the frequency of conflict is evidence of pacification because the same groups that fought one another prior to their absorption into a larger country did not do so subsequently.

2. Brecke himself raises this concern in his 1999 paper: "Obviously, the *sample* of cases for which data can be collected is significantly smaller than the population, particularly for conflicts in which the number of fatalities is towards the lower end of the range, for conflicts further back in time, and for parts of the world where written records are not readily available, especially for earlier times."
3. On the impact of small chance events on the course of history see Ferguson (2008) and Lebow (2010).
4. Only Goldstein (2011, 22), who flatly refuses to argue that reductions in war are part of an irreversible trend, concedes that, "with no doubt whatsoever, the comparison of the past fifty years with the previous fifty shows a dramatic decrease in warfare."
5. Roberts (2012). Nor was this an isolated incident. On September 26, 1983, Soviet Lt. Col. Stanislav Petrov opted not to report an alert indicating that five American nuclear missiles had been launched. He chose well: The false alarm was triggered by sunlight reflecting off of clouds. Had he reported it, as he was supposed to do, hundreds of millions would likely have died in the resulting exchange (Matthews, 2018). A 2014 Chatham House report (Lewis et al., 2014) documents no fewer than thirteen such nuclear near-misses.
6. It's possible to use different periods, of course: to assume that the probability of a systemic war breaking out in a given five-year period is 0.1, or that the probability in a given day is 0.000054757. Doing so makes little difference.
7. The probability of seeing k "successes" (wars) in n "trials" (years) given an underlying probability of success p is given by the probability mass function of a binomial random variable, $Pr(X = k) = \binom{n}{k} p^k (1 - p)^{(n-k)}$.
8. Using industry-standard 95% confidence intervals—single-sided, in this case, since the hypothesis is that p has fallen below 0.02.
9. As it turns out, much more complicated ones provide a very similar answer. Professor Aaron Clauset examined the same data I had, using a much more sophisticated technique, and concluded that it would take another 100 to 140 years to conclude that the Long Peace was more than a statistical anomaly (Clauset, 2018, 6).
10. Question 187 of Wave 6, carried out from 2010 to 2014. http://www.worldvaluessurvey.org/WVSOnline.jsp, accessed on 7/30/2014. Percentages reflect percent responding in the affirmative.
11. Carried out annually from 2003 to 2013, though some countries were only surveyed in later years. Source: http://trends.gmfus.org/transatlantic-trends-2013/, accessed on 8/2/2015. Percentages reflect percent responding strongly or somewhat in the affirmative.
12. Six countries (France, the United Kingdom, Italy, Portugal, Slovakia, and Bulgaria) appeared in the Transatlantic Trends survey but not in the World Values Survey (WVS). Their responses from 2012, the midpoint of WVS wave 6, were added to the WVS data. While the survey questions are identical and neither set of responses contains a neutral category, I should note that some of the countries in which both surveys were carried out exhibit differences of an

unsettling magnitude: the Turks are less bellicose by 10% in the Transatlantic Trends survey than in the WVS survey, while the Dutch are more bellicose by 13%. None of the conclusions about opinions in OECD vs. non-OECD countries change if those countries are omitted, however, so I have added those six countries' responses to the WVS responses.

13. A two-sample *t*-test produces a *p*-value of 0.946.
14. See Cook (1956) for details.
15. Technically, the CDC data register county-equivalents, not counties—the residents of Alaska live in boroughs and the residents of Louisiana live in parishes.
16. Goldstein (2011, 238) looks at the raw number of battle deaths per five-year period, while Pinker (2011*a*, 193–200) prefers to measure battle deaths as a percentage of world population.
17. Another point to bear in mind, one made well by Professor Tanisha Fazal (2014), is that battlefield medicine has made great strides over the years: a wounded soldier in the American Civil War could easily die of combat wounds that would not be fatal today. To some degree, then, a decrease in the number of battle deaths over time reflects the fact that the enemy soldiers we shoot are at a lower risk of dying from their wounds—not that we are less inclined to shoot them in the first place.
18. For more details see Cirillo and Taleb (2015).
19. Professor Eric Uslaner makes this point in an article whose title was good enough that I've borrowed it for this section (Uslaner, 1976).
20. It's true, of course, that, the risk of death to people outside of combatant countries is never *exactly* zero. Witness, for example, the accidental downing of Malaysia Airlines Flight 17 early in the present conflict between Russia and Ukraine. But the risk of being killed in that war for anyone outside of those two countries is about as close to zero as it gets. It's certainly not significant enough to include counting the entire world population as being at any significant risk. The overwhelming majority of the people reading this book, for example, would be very surprised to hear that their next-door neighbor had been done in by shrapnel from a 30mm Russian grenade launcher.
21. See, e.g., Campbell, Findley, and Kikuta (2017); Firchow and Ginty (2017); Kasten (2017).

Chapter 3

1. Holland (2014) is an excellent short introduction to the science of complexity.
2. Admittedly, international relations scholars have been relatively slow to recognize and grapple with the complexity of the international system. Some notable exceptions include Cederman (1997); DeNardo (1997); Epstein (1999); Richards (2000); Albert, Cederman, and Wendt (2010); Cudworth and Hobden (2010); and Bara (2014). While complexity science is highly mathematical in nature, some more historically oriented social scientists have taken an interest in it: see, e.g., Snyder and Jervis (1993) and Jervis (1997). For

an excellent agenda-setting piece for those working in the statistical tradition, see Chaudoin, Milner, and Pang (2015).

3. The term, now ubiquitous in the literature, comes from Ross (1977).
4. Early works include Langer and Gleason (1952), Adler (1957), and Jonas (1966), which tended on the whole to assume that isolationism sprang from dispositional causes. Later research (Wittkopf, 1990; Holsti, 1979) sought to tease out the sources of that disposition.
5. My own take on isolationism in this period, which describes it in far more detail and from which these numbers are drawn, has been published as Braumoeller (2010). Professor Joshua Kertzer (2013) provides an outstanding interactionist perspective on isolationism—one that explores the interaction of disposition and situation.
6. Arendt (2006, 32). The passage continues, in Eichmann's words: "I sensed I would have to live a leaderless and difficult individual life, I would receive no directives from anybody, no orders and commands would any longer be issued to me, no pertinent ordinances would be there to consult—in brief, a life never known before lay before me."
7. The specifics of the experiments that Milgram ran and their results are written up in Milgram (1963).
8. Some original video of the experiments is available on YouTube. It is utterly engrossing. I should note that Pinker himself mentions Milgram's experiments when discussing the ways in which ideology can lead to war (Pinker, 2011*a*, 558) but does not address the question of why the contextual pressure that Milgram documents would be less relevant to Enlightenment humanists.
9. See also Marlantes (2011, 32), where he writes of a different incident, in which he killed a North Vietnamese opponent who was about his son's age: "What is different between then and now is quite simply empathy. ... Back then, I was operating under some sort of psychological mechanism that allowed me to think of that teenager as 'the enemy.' I killed him... and we moved on. I doubt I could have killed him realizing he was like my own son. I'd have fallen apart."
10. For key details of this brief account see Jarausch (1969) and Clark (2013, 415–418). Clark does argue that the German strategy was not, strictly speaking, one of risk but rather a test of the threat represented by Russia (419); regardless of whether or not one thinks of such a test as constituting a strategy of risk, the same cannot be said with regard to the British.
11. The best-known discussion of the security dilemma remains Jervis (1978), though the term itself, or one very much like it (the "power and security dilemma"), comes from Herz (1951).
12. Indeed, his first book (Waltz, 1959) examined both, as well as the systemic influence of anarchy, and his second (Waltz, 1967) was a work on the foreign policies of democracies.
13. Elias (1988, 180–181), cited in Landini and Dépelteau (2017, 19).
14. In fairness, Professor Pinker does mention anarchy on page 291 of *Better Angels*. He notes that realism, the international relations paradigm most prominently

associated with anarchy, is sometimes defended as a consequence of human nature and that that understanding of human nature is fundamentally flawed. The critique very much misses the point: Waltz's emphasis on the importance of international anarchy was a reaction *against* the so-called human-nature realists who had come before.

15. See Hanlon (2018) and Bell (2018) for thoughtful examples.
16. The extent to which the more liberal Enlightenment thinkers themselves were willing to countenance just war is an interesting question. Orend (2000), for example, argues that Kant believed in the possibility of a just war, while Williams (2012) disagrees, arguing that Kant's clear differences with just war theorists like Hugo Grotius, Samuel von Pufendorf, and Emer de Vattel in *Perpetual Peace* put him squarely in the anti–just war camp.
17. Season 6, episode 8 ("No One"). Pinker himself makes the same point in less colorful language: "Individual people have no shortage of selfish motives for violence. But the really big body counts in history pile up when a large number of people carry out a motive that transcends any one of them: an ideology" (Pinker, 2011*a*, 556)
18. International Coalition for the Responsibility to Protect (2017).
19. On this point see, e.g., Janik (2017).
20. See https://www.bloomberg.com/billionaires/profiles/william-h-gates/.
21. The small blip in the upper-left corner is not an illusion: some households reported negative net household income.
22. For more information about these datasets as well as links to download pages, visit http://www.census.gov/programs-surveys/acs/technical-documentation/pums.html.
23. Cirillo and Taleb (2016*a*); Clauset (2018). Two very recent rejoinders (Spagat and van Weezel, 2018; Cunen, Hjort and Nygård, 2018) are worth noting but, at this writing, have yet to pass through peer review.

Chapter 4

1. For these analyses, I utilize Version 4.0 of the MID dataset (Palmer et al., 2015). Including the modifications proposed by Gibler, Miller, and Little (2016) produced no substantive changes in the results.
2. http://abcnews.go.com/GMA/video/north-korea-threatens-final-doom-us-28719922
3. For a thoughtful dissent from this position see Palmer et al. (2015, 236–238). While their concern about selection on the independent variable is valid in many cases, it should represent less of a problem in the present study in that I utilize no independent variables.
4. For a detailed discussion, see Sarkees (n.d.).
5. Braumoeller and Carson (2011). The Braumoeller-Carson measure of political relevance can be generated quite easily using variables from the Correlates of War dataset. To do so, I used their formula and estimated coefficients: $\Lambda(4.801 + 4.50 \times \text{contiguity} - 1.051 \times \log(\text{distance}) + 2.901 \times \text{major power})$,

where Λ denotes a standard logistic function, defined as $f(x) = 1/(1 + e^{-x})$, and contiguity, distance, and major power status are defined as in Maoz and Russett (1993).

6. I used a standard receiver-operator curve (ROC) to gauge the ability of each measure to categorize uses and non-uses of force, as measured by the MID data. The area under the curve (AUC) favored the Braumoeller-Carson measure (0.90) over the Maoz-Russett measure (0.85). I also generated a hybrid measure, in which the Braumoeller-Carson distance adjustment remained the same, but in accordance with the Maoz-Russett model, all major powers were considered to be relevant to all other countries in the system. It outperformed the Maoz-Russett model, but just barely (0.86). Yet another measure, devised by Quackenbush (2006), includes alliances as indicators of political relevance, but doing so miscategorizes so many politically irrelevant states as relevant that the measure is the worst performer of the four (0.81).
7. Summing the scores across all pairs of states does assume that, under otherwise identical circumstances, a pair of countries with a political relevance score of 1 is twice as likely to fight as a pair of countries with a political relevance score of 0.5. Fortunately, that assumption is built into the conflict model that generated the political relevance scores, so it should be justified here.
8. See James and Matteson (2014) for details. If you want to know more about how this technique works, you can find an intuitive explanation in the vignette titled "The Mysterious Case of the Vanishing Frontrunner," in the Online Appendix.
9. Starting on page 225 of the Appendix, I argue that that's a very debatable claim.
10. It's possible to make it sensitive to changes over even shorter periods. I didn't do so because I wanted to avoid having it pick up transitory spikes or dips that were unrepresentative of a period as a whole. At the same time, I wanted it to be flexible enough to pick up changes when they did occur. A decade strikes a good balance between those two objectives.
11. See http://cow.dss.ucdavis.edu/data-sets/MIDs.
12. http://www.worldstatesmen.org/India_princes_A-J.html.
13. We have to use wars, for the simple reason that militarized disputes with non-state entities were not recorded, so the results are necessarily less granular.

Chapter 5

1. But see Levy (1983, 83) for arguments to the contrary.
2. See, e.g., Murray et al. (2002) and Ghobarah, Huth, and Russett (2003).
3. While the ratio of military to civilian deaths obviously varies from one war to another, the average ratio has remained surprisingly constant, at roughly 1:1, over the course of three centuries (Eckhardt, 1989).
4. For an overview and summary, see Geller and Singer (1998).
5. Technically, its results can't be guaranteed when the mean of a distribution is undefined or infinite, and as it turns out that's the case here. So while I don't rely too heavily on the results, I include them for the sake of thoroughness.

6. Cederman, Warren, and Sornette (2011, 622–623). The authors rely on the asymptotic properties of their estimator of the slope coefficient when testing for differences in slopes, while I use Monte Carlo simulations to obtain a bootstrapped estimate of the distribution of differences. The preference for one test versus another is akin to the preference for chocolate ice cream versus vanilla—mostly just a matter of taste.
7. Professors Jack Levy and William R. Thompson note this trend and also conclude that the frequency of war onset was lower in the post-Napoleonic period than it had been previously (Levy and Thompson, 2011).
8. At the time of this writing, it is being reported that Paraguay has at long last managed to locate its first oil reserves in the Chaco region, though environmental activists are seeking to prevent its extraction.
9. Technical note: I also adjust the output to take into account multiple inference, using the method of Holm (1979). The standard criterion for statistical significance is a result that would only have happened by chance one time out of twenty. If you run twenty tests using that same criterion, you're very likely to get a false positive result (that is, a result that looks statistically significant but is really just the result of chance). Randall Munroe offers a funny and brilliantly simple explanation of this point at http://xkcd.com/882/.
10. I tried to find a test that would give me the answer to this question, but the statistics of power-law distributions is a recent topic of relatively narrow interest. In the end, I derived a simple test myself, from first principles, and proved to myself using simulated data that it provides reasonable answers. I've laid it out in the Appendix, starting on page 238, along with code to reproduce it in the R programming language.
11. The easiest way to see this is to set $\alpha = 2$ and note that the denominator equals zero; lower values of α just make an infinite outcome more likely. Aaron Clauset's clear primer at http://tuvalu.santafe.edu/aaronc/courses/7000/csci 7000-001-2011-L2.pdf is a useful reference.
12. Cirillo and Taleb (2016*b*) were, to my knowledge, the first authors both to note this problem and to offer a mathematical solution. Their solution has no impact on the substantive results in this chapter.
13. Technically, the odds of complete annihilation are even better than that, since the data only reflect battle deaths and therefore understate total deaths from war, but it's very hard to know by how much they do so.
14. The scattered red points in the top plot are worth noting as well: They indicate that the battle deaths per thousand world population measure, unlike the other two measures, doesn't conform especially well to a power-law distribution. I continue to discuss its escalatory properties for the simple reason that the numerator *does* conform to a power-law distribution.
15. Correlates of War data. This period of relative stability isn't a fluke: While there are obviously fluctuations from time to time, the average size of a country's military forces in 2000 is about the same as it was in 1880.
16. I use these rather than onset dates to be as fair as possible to the decline-of-war thesis. If attitudes toward killing determine the lethality of wars, it'd be best to

average those attitudes across the course of the war, rather than measure them at the start or at the finish. Time is just a proxy.

17. The data, again, are from the Correlates of War project. The lines are calculated via ordinary least squares. The clear lower bound on the data is the result of the definition of what constitutes a war: No conflicts with fewer than a thousand battle deaths are counted, which means that the intensity threshold is higher for small states than it is for large ones. Nothing *mandates* that the conflicts above that line have a negative slope, however: If larger states systematically fought more intense wars than smaller states do, the lines would be sloped upward.
18. I'm not too concerned about the impact of slower-growing militaries on the findings about war intensity, for a couple of reasons. First, it's less of a problem. The countries that fight wars tend not to be the countries with the highest population growth in the postwar period, so while militaries grow more slowly than populations, they grow a *lot* more slowly than world population. Second, if it were a problem, it would result in a spurious positive result, and none is evident (yet). What it does mean, I think, is that we have no really perfect measure of the deadliness of war. Severity ignores the growth in population, while intensity and prevalence bake in a bias toward false positives. We can't use war size as a percentage of military size as a measure because militaries get bigger when wars get bigger. Intensity, I think, provides the best answer we have at this time. There's probably a better adjustment that lies somewhere between no adjustment and dividing by pooled population size, but figuring out what it would be is a much harder problem than it might seem. As long as intensity and severity give us the same answer, it doesn't make a lot of sense to let some notional best somewhere in between them be the enemy of the good.
19. This scenario is a variant of a common problem from courses on basic probability, the *two-sided gambler's ruin.*
20. We actually know surprisingly little about why wars get as big, and as costly, as they do. Very few studies have focused on the power-law distribution of war outcomes, and even fewer have offered a compelling explanation for it (though see Cederman [2003] and Pinker [2011*a*, 220]).

Chapter 6

1. Technically, it's possible that conflict was on the rise during that time because the number of things to fight over (borders for newly independent states, independence for national groups) grew over the same period. And if the number of things to fight over grew quickly enough, it might even outpace the rate at which people were becoming less inclined to fight over them (if in fact they were). Under those circumstances, the decline-of-war theorists could theoretically still be right about changes in human nature even if they were wrong about trends in conflict.
2. By sheer coincidence, as I was writing this chapter Ms. Libby Jenke and Professor Christopher Gelpi (2017) were conducting a study along the same

lines, taking as a canonical model Bennett and Stam's 2003 behavioral model of war. If I were going to choose one model to reexamine, that incredibly comprehensive model would be a prime candidate. Jenke and Gelpi find, as I do, that there is significant and underappreciated variation in the impact of war-causing variables over time. More relevant to the question being explored here, their results indicate no systematic upward or downward trends in the potency of the variables that they examine.

3. Why am I only examining international conflict? The short answer is that the overwhelming majority of quantitative studies during this period are studies of international conflict (and more specifically, studies of militarized interstate disputes). Studies of extrastate and nonstate warfare are basically nonexistent, and those studies of civil war that do exist are generally of sufficiently limited temporal scope that a long-term trend would be difficult to capture.
4. Specifically, they examined the *American Political Science Review, American Journal of Political Science, Journal of Politics, European Journal of International Relations, British Journal of Political Science, World Politics, International Security, International Organization, Journal of Conflict Resolution, Journal of Peace Research, International Studies Quarterly, International Studies Review, Conflict Management and Peace Science,* and *International Interactions.* The criteria were: 1. the dependent variable is war, militarized interstate dispute (MID; Correlates of War project), or armed conflict (UCDP/PRIO data); 2. the temporal scope extends backward at least to 1945; and 3. the statistical model utilized is logit or probit. I am very grateful to David Nield, Steven Mahr, Anna Meyerrose, Daniel Kent, Austin Knuppe, and Miguel Garza Casado for the time and effort that they put into these replications and extensions.
5. We actually started out using structured additive regression (STAR) models, following an excellent paper by Professors Gregory J. Wawro and Ira Katznelson (2014). We soon switched to generalized additive models (GAMs; see Hastie and Tibshirani 1990) when we found the STAR algorithm to be somewhat fickle for harder optimization problems. When both converged, they gave essentially identical answers. Analysis was done in R with the mcgv and R2BayesX libraries, respectively.
6. This a monadic rather than a dyadic study, meaning that it measures the militarized dispute involvement of individual countries rather than pairs of countries.
7. *Shameless plug alert!* See Braumoeller et al. (2018) for one example of such a model.
8. When I visited Vanderbilt University to talk about this project, Professor Brenton Kenkel raised this point about these findings. He was right to do so. I include them, with apologies to Brenton, because at the present time they're the best estimates we've got. I hope they won't be ten or fifteen years down the line.

Chapter 7

1. See, e.g., Tilly (1992). Other areas of the world took very different paths, as one might expect; see Suzuki, Zhang, and Quirk (2016); Phillips and Sharman (2015).
2. An organization that actually runs the world could, one hopes, meet somewhere more posh than the Westfield Marriott in Chantilly, VA, which hosted the Bilderberg conference from June 1 to 4, 2017. (Or perhaps that's what they want us to think.)
3. While the concept of international (or sometimes world) order has been a fairly common theme in writings by historians, policymakers, classical international relations scholars, and political scientists of the so-called English School (Deutsch et al., 1957; Morgenthau, 1960; Bull, 1977; Knock, 1992; Finer, 1997; Adler and Barnett, 1998*a*; Howard, 2000; Slaughter, 2004; Fukuyama, 2011, 2014; Kissinger, 2014), it has only recently become prominent in the mainstream international relations literature. Since the publication of Kenneth Waltz's *Theory of International Politics* in 1979, that literature has been grounded in the assumption that the international realm is anarchic, in that there is no meaningful international order or hierarchy of authority. As Professor David Lake (2014, 61) puts it, "[t]here is no clearer example of how paradigmatic assumptions blind scholars to reality" than the discipline's decades-long fixation on anarchy. Lake, along with Professor Ayşe Zarakol, are at the forefront of a movement to replace anarchy with hierarchy in our understanding of international politics (Bially Mattern and Zarakol, 2016); their books *Hierarchy in International Relations* and *Hierarchies in World Politics* should be required reading for serious students of international relations. The revival of the study of international order owes much to the work of Professors G. John Ikenberry (2001; 2011; 2014*b*) and Robert Gilpin (1981), as well as to a much older literature on political order more generally (e.g., Huntington [1968], which foreshadowed Fukuyama [2011] and Fukuyama [2014]; Finer [1997] is also well worth seeking out). While it obviously bears on the question of Great Power war, the study of international order has yet to make real inroads into the general study of international conflict—although see McDonald (2015) and Warren (2016) for promising recent exceptions.
4. See footnote 7, this chapter.
5. See also Tilly (1992, ch. 3). Cohen, Brown, and Organski (1981) extend this argument to cover the role of violence in the formation of new non-European states in the postwar period.
6. See, e.g., Pinker (2011*a*, 288–294).
7. For a similar argument see the work of Professor Peter Turchin (2007, 2016). Professor Kalevi Holsti (1991) argues in a similar vein that international orders often contain the seeds of their own demise. I feel compelled to note that Elias's last sentence ends, "until, with a further integration, a still greater reduction of distances, they too gradually grow together and world society is pacified." We'll see.

8. See, e.g., Spruyt (1994); Caporaso (2000).
9. See, e.g., Clark (1989, 14).
10. To extend the analogy of books arranged alphabetically on a shelf, a purely anarchic system might resemble the jumble of books that's piled on my office desk as I write this. It's arguably in *some* sort of order, in that I can often find one of the books in it fairly quickly. But it's also in a constant state of flux, as I move books here and there, and I can spend a fair bit of time digging through it before I find the book I'm looking for. Any order that might characterize it, in other words, is accidental and transitory. This is the condition that sorting books by the author's last name, as I've done on my bookshelves, is meant to solve.
11. In the case of Utrecht, the major European powers agreed that no single power should be allowed to grow strong enough to threaten the rest, a premise that justified the stipulation that the French and Spanish crowns should never be united. On this point, Schweller (2016) is especially helpful.
12. I owe this concise formulation to Lake (2009, 52). Professor Lake used it to describe security hierarchies, but it applies equally well to security regimes more generally.
13. These requirements mean that regional orders, such as the one comprising the countries of the Association of Southeast Asian Nations, or ASEAN, are not included. While excellent work has been done on regional orders (see, especially, Lake and Morgan [1997]; Solingen [1998]; Stewart-Ingersoll and Frazier [2012]), it's not at all clear to me how small a region can be (Stewart-Ingersoll and Frazier identify twelve, some of which are quite small) and, accordingly, where one draws the line between regional orders and alliances. I'm grateful to Professor Brett Ashley Leeds for a conversation that highlighted these conceptual challenges.
14. This definition is a mashup of a classic definition of domestic political order from political scientist Seymour Martin Lipset (1959, 86) ("the capacity of a political system to engender and maintain the belief that existing political institutions are the most appropriate or proper ones for society") and, less obviously, one of international relations scholar Hans Morgenthau's (1948, 29) sources of political power, "the respect or love for men or institutions."
15. For much more detailed overviews of the construction of the Concert, see Kissinger (1957), Schroeder (1994, ch. 12), and Jarrett (2013, ch. 4).
16. See Hurd (1999).
17. Howard (2000, 40–45, 67–71, 74–76). To be clear, the meanings of the terms "conservative" and "liberal" in this context are quite different from their meanings in modern American politics.
18. See Kant's essay *Perpetual Peace: A Philosophical Sketch* (1795), reprinted more recently as Kant (2003).
19. Burke (1790); Deutsch et al. (1957). Deutsch and coauthors, who originated the phrase, make a distinction between *pluralistic* security communities, in which the members of the community retain their legal independence, and *amalgamated* security communities, in which they do not. The United States

is an example of an amalgamated security community, as is Italy. In the present context, I focus specifically on pluralistic security communities. See also Adler and Barnett (1998*a*).

20. See Keegan (1993, 24–28) for proof that I am not making this up.
21. The backstory provides useful context: At the time, nutmeg was thought to cure the Plague. See Milton (2000) for details.
22. Rosenau (1966) and Zimmerman (1973) deserve credit for theorizing about the general nature of issues and their role in foreign policy. Mansbach and Vasquez (1981) and, later, Diehl (1992) became the standard-bearers for an issue-based paradigm in international relations research. Hensel (2001) introduced a new project, the *Issues Correlates of War* (ICOW) dataset, which continues to expand from territorial disputes to riperian and maritime clashes to identity claims.
23. The realists and neoclassical realists focus most narrowly on the importance of power in international politics. See Waltz (1979), Organski and Kugler (1980), Gilpin (1981), and Mearsheimer (2001) for prominent works in the former vein and Wohlforth (1993), Christensen (1997), and Schweller (1998) for some outstanding examples in the latter.
24. Huntington (1958), Richardson (1960*a*), and Jervis (1976, ch. 3) are foundational, as is Wallace (1982) in the empirical literature (though mostly for having had its findings called into question by some of the work that followed); see also Diehl (1983), Williams and McGinnis (1988), Morrow (1993), Jervis (1993), Kydd (1997), Sample (1997), Diehl and Crescenzi (1998), Kydd (2000), and Rider (2009).
25. See Hayes (1931) and Kohn (1944) for classic early works and Anderson (1987) for a more recent benchmark. More recent works include Snyder (1993), Van Evera (1994), Kaufmann (1996), Braumoeller (1997), Davis and Moore (1997), Fearon and Laitin (2003), Laitin (2004), Chapman and Roeder (2007), Cederman and Girardin (2007), Bhavnani and Miodownik (2008), Cederman, Buhaug, and Rod (2009), Cederman, Wimmer, and Min (2009), Cederman et al. (2013), and Sambanis and Shayo (2013). Hall and Malešević (2013) present a useful survey of recent research.
26. Here, Russett (1993) and Bueno de Mesquita et al. (2003) are widely cited. The question of the connection between political democracy and the absence of conflict consumed the empirical international relations literature throughout the 1990s and continues to be hashed out sporadically. Especially interesting works include Lake (1992), Russett (1993), Maoz and Russett (1993), Rousseau et al. (1996), Braumoeller (1997), Schultz (1999), Cederman (2001), Huth and Allee (2002), Dafoe, Oneal, and Russett (2013), and Tomz and Weeks (2013).
27. See especially Wolford (2007), Haynes (2012), and Croco and Gartner (2014).
28. Here, the argument can be traced back to the late eighteenth and early nineteenth centuries, when classical liberals such as Adam Smith and Cannan (1994) and Richard Cobden (1997) argued that commerce would be (in Cobden's words) the Grand Panacea, which would ultimately produce peace among nations. More recent empirical literature on the subject includes

Barbieri (1996), Oneal and Russet (1997), Gartzke, Li, and Boehmer (2001), Mansfield and Pollins (2001), and Russett and Oneal (2001); Mansfield and Pollins (2006) is a useful review and summary.

29. See *inter alia* Vasquez (1995), Hensel and Mitchell (2005), Goddard (2006), Gibler (2007), Fravel (2008), Owsiak (2012), and Johnson and Toft (2014).
30. This literature is mostly quite recent: see, e.g., Hendrix and Glaser (2007), Salehyan (2008), Buhaug, Gleditsch, and Theisen (2010), Buhaug (2010), Butler and Gates (2012), Raleigh and Kniveton (2012), Tir and Stinnett (2012), Hendrix and Salehyan (2012), Busby et al. (2013), Salehyan and Hendrix (2014), and Jones, Mattiacci, and Braumoeller (2017).
31. For details see Smith (2011) and the United Nations summary at http://www.un.org/en/peacekeeping/missions/unoci/elections.shtml.
32. Adler and Barnett (1998*b*, 49–56) make this point in the context of security communities.
33. Professor Lars-Erik Cederman (2001) makes the case for a normative learning process of this nature in democratic states.
34. See, e.g., Judt (2005, 143–144).
35. Professor Lake (2009, 105–111) makes this point well.
36. See Americas Watch Committee (U.S.) (1987).
37. Most notably, see von Clausewitz 1976 [1832]; see also Pillar (1983) and Werner (1998).
38. Von Clausewitz (1976 [1832], 69). The full context of the quote is worth dwelling on. It comes from a note dated July 10, 1827, in which Clausewitz briefly described his plans for revising the text of *On War.* In that brief note, he stated in an aside that "If an early death should terminate my work, what I have written so far would, of course only deserve to be called a shapeless mass of ideas." Precisely such a death overtook him four years later, leaving all but the first chapter of the first book unrevised. In the 1827 note, however, Clausewitz underscores the importance of his understanding of war as the continuation of politics, stating that "[i]f this is firmly kept in mind throughout it will greatly facilitate the study of the subject and the whole will be easier to analyze."
39. See Hastings and Jenkins (1983) for details.
40. Worthwhile early discussions of the impact of international regimes and institutions on information include Keohane (1984), Keohane and Martin (1995), and Oye (1986).
41. That is, the five permanent members of the United Nations Security Council—China, France, Russia, the United Kingdom, and the United States—plus Germany.
42. Gaddis (1997, 167–176) contains a useful quick summary of the crisis. On the value of shared "national culture" for reducing uncertainty see Deutsch et al. (1957, 56–57); on the impact of legitimation in hierarchies see Lake (2009, 94–96), and in the specific case of the United Nations see Claude (1966), Voeten (2005), and Westra (2010).
43. See, e.g., Boehmer, Gartzke, and Nordstrom (2004), Bearce, Flanagan, and Floros (2006), and Shannon, Morey, and Boehmke (2010).

44. The need to maintain some ambiguity in alliances was described by Professor Glenn Snyder (1984). Strong allies can also serve as mediators to help defuse crises, but current scholarship is somewhat divided on how effective they are in that capacity. See Fang, Johnson, and Leeds (2014) for an optimistic assessment. Gelpi (1999), by contrast, finds that stronger states with alliance ties to a potential disputant are *less* effective than weaker, more neutral states are at mediation.
45. On China's security guarantee and the uncertainty that it creates, see, e.g., Bajoria (2008) and Raska (2014).
46. Shannon (2009). Moreover, Shannon, Morey, and Boehmke (2010) demonstrate that such institutions also decrease the duration of conflict when it does break out.

Chapter 8

1. General diplomatic histories are, alas, going out of style: the welcome realization that the history of the world is more than just the history of dead white men has produced a turn toward a richer and more diverse set of subjects. That said, there is quite a bit of material for anyone seeking a more in-depth understanding of Great Power politics in the nineteenth and twentieth centuries. The indispensable starting point is Paul Schroeder's *The Transformation of European Politics*, which chronicles the underappreciated transition from balance-of-power politics to collective international governance following the Napoleonic Wars. The old Harper & Row series entitled *The Rise of Modern Europe,* comprising twenty volumes by prominent historians and spanning seven centuries, is a treasure if it can be found in the dusty shelves of a used bookstore. René Albrecht-Carrié's *Diplomatic History of Europe Since the Congress of Vienna* , a standard for many years, is a bit long in the tooth, as is A. J. P. Taylor's *Struggle for Mastery in Europe*, though both are still useful. F. R. Bridge and Roger Bullen's *The Great Powers and the European States System 1814–1914* remains a remarkably accessible history of the long nineteenth century as a whole. Turning to the twentieth, Sir Harold George Nicolson's *The Congress of Vienna* is an insightful study of one international congress by a man who attended another (Nicolson served in the British Foreign Office and attended the Paris Peace Conference in 1919; in that same vein, see also Henry Kissinger's *A World Restored*). Gordon Craig and Alexander George's *Force and Statecraft,* a too-rare collaboration between an historian and a political scientist, is a succinct and thoughtful gem of a book: the chapter on the nineteenth century especially stands out. William Langer's *European Alliances and Alignments 1871–1890* is a captivating study of Bismarck's breathtaking diplomacy. E. H. Carr's *The Twenty Years' Crisis,* although its thesis is debatable, should be required reading on the interwar period, and John Lewis Gaddis's *The Cold War: A New History* is a reasonably comprehensive history written for a general audience. Geoffrey Blainey's *A Short History of the Twentieth* Century is lucid and concise, while Martin Gilbert's *A History of the Twentieth Century*

is at least lucid, though the author should be commended for taking on the painful task of cutting his original 2,700-page three-volume epic down to a more manageable and remarkably rich eight hundred.

2. See, e.g., Taylor (1954, xix–xxxvi) and Artz (1934). Interestingly, Schroeder (1994, 673) argues that the post-Vienna Powers were actually quite lax when it came to policing revolution.
3. Two examples nicely illustrate the magnitude of this confusion: while Bridge (1979) argues that the Concert ceased to operate in 1823, Jarrett (2013, 369) argues that it only functioned *after* 1823.
4. See Elrod (1976, 163), Jervis (1985, 58), and Schroeder (1994, chs. 15–17).
5. Metternich, who groused that the British would not accept French mediation in the event of an Irish uprising, refused to take part, and Prussia followed Austria's lead. See Brewer (2001, 256–257).
6. See Binkley (1963, 163) for details.
7. Even more confusingly, the Congress of Berlin (1878) is sometimes cited as the last Congress of the Concert system, but by that time the Concert existed mainly as an ideal-typical aspiration rather than as a functioning system, as William Gladstone's attempts to revive it during the Near Eastern Crisis of the midto late 1870s demonstrates (see, e.g., Medlicott [1969], as well as Jenkins [1997, 409, 501]).
8. On this point see Bass (2008).
9. See Mitzen (2013*b*, 41–45 and *passim*).
10. Schroeder (1994, ch. 12) makes the case for the post-Vienna territorial settlement as an important element in the working of the Concert, while Slantchev (2005) argues that it was in and of itself sufficient for peace.
11. The Third, and final, Italian War of Independence arose in the context of the second of these.
12. This section and the next are drawn, with some modifications, from chapter 3 of Braumoeller (1998).
13. Eyck (1964) and Taylor (1967) are classic portraits of Bismarck, updated usefully by Steinberg (2011). Taylor (1954) and Albrecht-Carrié (1958) usefully situate his policies in the larger context of European politics.
14. As Professor David Calleo (1978, 10) put it, "Essentially, Bismarck's celebrated foreign policy consisted of a complex set of agreements meant to keep all the other powers perpetually off balance."
15. The details and chronology in this section follow Albrecht-Carrié (1958, 164–186).
16. Bismarck intentionally created an atmosphere of ill-will between Alexander II and William I that prompted a summit between the two. As Taylor (1954, 260–261) writes of the meeting, "[i]t served no purpose. Bismarck had cast tsar and emperor for a quarrel, and they had to quarrel to their mutual bewilderment, until the alliance with Austria-Hungary was safely concluded."
17. See Albrecht-Carrié (1958, 178–179) for details.
18. See Joll (1992, 46–47); Albrecht-Carrié (1958, 178). Dissenting on the third of these points is Taylor (1954, 267).

19. McDonald and Rosecrance (1985, 71). Healy and Stein (1973, 49), an earlier study that inspired the analysis in question, points out that balance in the sense of polarization—the logical result of an equilibrist balance-of-power policy—can be undesirable in the same way that certain game-theoretic equilibria can be undesirable: The mere fact that they exist and are stable doesn't make them desirable. States will go to great lengths to avoid "balanced" situations, such as those which precede wars.
20. See Leeds (2003) for more comprehensive evidence that information provision in defensive alliances promotes peace.
21. Eyck (1964, 289–293). On the dangers of secret alliances see Bas and Schub (2016).
22. Kennedy (1980), a classic work on the subject, makes the case that it would have been exceptionally difficult.
23. On this and the following paragraph, see Albrecht-Carrié (1958, 208–214).
24. These are just a few of the quotations collected in Fischer (1975, ch. 3).
25. The phrase is from Ensor (1936, 259).
26. The self-justifying nature of the German action has been noted in a variety of histories on the subject, including Taylor (1954, 375) and Albrecht-Carrié (1958, 226).
27. For an excellent summary of the intricacies of this process, see Berghahn (1973, ch. 2).
28. Craig (1978, 321–326); figures are from Kennedy (1987, 203). Specifically, the tonnage of the British fleet was 1,065,000 in 1900, 2,174,000 in 1910, and 2,714,000 in 1914; for Germany, the figures are 285,000 in 1900, 964,000 in 1910, and 1,035,000 in 1914.
29. It had not escaped Schlieffen's notice that Belgium's German border was fortified but her French border was not. See Craig (1978, 317, fn. 48).
30. Herrmann (1996) is an outstanding exploration of the impact of changes in the strength of European land armies on decisionmakers' calculations. For a skeptical view of "windows of opportunity" arguments, see Lebow (1984).
31. Political scientists are divided on the question of whether multipolar systems are inherently more susceptible to conflict. On the pro side, see, e.g., Waltz (1964) and Mearsheimer (1990); for arguments against, see Hopf (1991) and Van Evera (1990). My own take is that, while other factors may be more influential, the increased uncertainty associated with multipolarity cannot improve the prospects for peace.
32. Clark (2013, 417–418, ch. 12).
33. For details of this remarkable gambit, see McMeekin (2013, 274–278). To the immense relief of Admiral Tirpitz, the Kaiser rejected the plan.
34. Keegan (1999, 59–60).
35. Article 16 obliged member states to apply sanctions in the event of aggression and gave the Council the ability to recommend an armed response, but no one was obligated to follow such a recommendation. See Luard (1998, 5).
36. For a comprehensive and authoritative history of Soviet foreign policy in the early years, see Ulam (1974); see also Gorodetsky (1994), Kokoshin (1998),

Zimmerman (1969), Lynch (1987), and Nation (1992) on the sources of Soviet foreign policy.

37. For an outstanding history of Japan during this period, see Toland (2003); for a good and succinct international history, see Iriye (1987). On the racial component of Japanese policy, see Dower (1986).
38. Howard (2000, 68, 71). See Burleigh (2001) for a general history and Jäckel (1972) for a focused discussion of Hitler's worldview.
39. For a detailed discussion of the evolution of American perceptions, see Lynch (1987).
40. Lenin (1939). The Maisky-Litvinov communiqués which culminated in Maisky's January 1944 letter to Molotov are a rare glimpse into Soviet thinking on this subject. The letter began with the premise that ideology would play a role in the postwar period, *but only among imperialist powers;* he predicted a coming clash between the United States and Britain over colonies and trade. The Soviet Union would be free to engage in precisely the same kind of balancing behavior that Britain had become known for in the eighteenth century: playing one side off of the other. The goal, however, was a breathing space which would eventually (in thirty to fifty years, by Maisky's estimate) permit the conversion of Europe to socialism. For the details of this communiqué, which was preserved in the foreign policy archives of what is now the Russian Federation, see Zubok and Pleshakov (1996, 28–30).
41. While the distinction is not totally clean, in broad strokes the leaders of the Soviet revolution emphasized fomenting revolt in advanced Western democracies while Chinese Communists were more interested in decolonization and the development of fraternal states in the Third World. As a result, two Communist powers with different beliefs about the most important priorities for the world Communist movement often fought for leadership of that movement, developing rival networks of clients that undermined international socialist unity. For an excellent account of this rivalry, see Friedman (2015).
42. Soviet obstreperousness did prompt the passage of the Uniting for Peace Resolution in 1950, which allowed the General Assembly to take action to preserve the peace if the Security Council were deadlocked. The only time that that authority has been used to recommend military action, however, was in the Korean War. See Carswell (2013) for a discussion of the history of the resolution.
43. In an early quantitative study, for example, Wilkenfeld and Brecher (1989, 164–165) found that the U.N. became involved in 91% of all wars in their sample but only involved itself in 35% of minor clashes.
44. Before it was widely recognized, this problem led to some puzzling conclusions. Diehl, Reifschneider, and Hensel (1996), for example, examined the long-term impact of United Nations involvement in crises and found that UN involvement in the cases in their sample *increased* the probability that a crisis would recur in the next ten years, from 47.3% to 56.9%—though given the number of cases in the study, a difference that small could easily have been

the result of chance. For the same reason, the net effect of peacekeeping more generally is very difficult to calculate (Fortna, 2013, 568).

45. I am grateful to Mr. Andrew Goodhart for his work on this part of the project and for thought-provoking discussions about international order more generally.
46. For example, Israel, which met neither criterion in 1995, ranks below Macedonia, and Panama, which still housed substantial numbers of American troops entirely because the Canal Zone had not yet been completely returned to them, ranks above Germany and Japan. See Lake (2009, 78–79) for details.
47. Specifically, I bootstrapped 2,500 resamples from each sample and measured the 2.5% and 97.5% quantiles of the distribution of their estimated means.
48. The wide uncertainty bars are largely the result of the relatively small sample of pairs of cases in this subset. In a given year with n League members, there are $\frac{n(n-1)}{2}$ possible pairs of League states, but there would only be $3n$ pairs that combine a League state with either Germany, Japan, or the Soviet Union. That's also why I didn't include all possible pairings of international orders during this period—the sample sizes would have been smaller still.

Chapter 9

1. It is instructive to note that the Soviets understood human rights in a very different way: their focus was on rights such as the right to education, medical care, shelter, and jobs. See Dean (1980).
2. To my mind, no one makes this point more convincingly than Robert Axelrod (1984) in his most famous book, *The Evolution of Cooperation.*
3. My colleague, Professor Randy Schweller, makes this point well; see Schweller (2014).
4. The debate, between Niall Ferguson (pro) and Fareed Zakaria (con), barely moved the needle. By the end, only 71% believed that the liberal order was still thriving.

Appendix

1. For a dissenting view, see Cunningham (2013).
2. The counts differ because the threshold for conflict differs, as mentioned above.
3. I haven't done the same for other kinds of wars, mostly because the trends in civil conflict dominate the data. The small number of wars in other categories make trends across these four categories much harder to spot.
4. Because the official Correlates of War system membership database has not been updated as frequently, I extended the 2011 numbers through 2014; the estimate of the number of conflict opportunities might therefore be ever so slightly off in the last three years, but certainly not so much as to alter our conclusions.

5. These results confirm Fearon and Laitin's (2003, 77) and Hegre's (2004, 243–244) observations that the bulge in the original graph is due to an accumulation of civil wars rather than any increase in initiations.
6. I'm very grateful to Professor Kristian Skrede Gleditsch of the University of Essex for bringing this point to my attention at a conference in the summer of 2016 and for elaborating on it later via e-mail (personal communication, June 16, 2016).
7. http://polisci.emory.edu/faculty/dreiter/JCRRSHappendix10-15-14.pdf, accessed December 7, 2015.
8. The authors note in their abstract that "in more than 30 percent of the ninety-five COW interstate wars, codings of at least one of the key variables...needs to be revised."
9. Another modification to the COW dataset (Gleditsch and Ward, 1999; Gleditsch, 2004) deserves mention. It is based on a different definition of statehood and a different list of members of the interstate system. Unfortunately, the data are not dyadic, so we can't tell how many non-wars there are, and it doesn't contain contiguity or distance information, so I can't calculate political relevance. In short, there's no way to use these data in their present state to calculate the rate of conflict initiation, as I have done above.
10. More specifically, the coefficient on year, while negative, is substantively very small (−0.001783) and fails to reach statistical significance (s.e. = 0.002390, $p = 0.457$). The dispersion parameter indicates that the variance is roughly twenty times the size of the mean, so relaxing the assumption that the two are equal is a very good idea.
11. Adding $year^3$ to the equation renders all of the coefficients statistically insignificant.
12. War intensity provides a "moderately" good fit, according to the authors: it passes the first test and outperforms the exponential distribution, though its performance relative to the other two distributions tested is ambiguous.
13. After I had finished the analysis, I became aware of the technique laid out in Cirillo and Taleb (2016*a*), which takes a different approach than mine does. Because the authors came to the same conclusion that I did about the decline-of-war thesis and their article had already been accepted for publication, I didn't bother to reproduce my own results with their technique.
14. Out of curiosity I explored the normal approximation as well as basic and BCa confidence intervals, and none outperformed the percentile confidence intervals.

BIBLIOGRAPHY

Adler, Emanuel and Michael N. Barnett. 1998*a*. A Framework for the Study of Security Communities. In *Security Communities*, 3–65. Number 62 *in* "Cambridge Studies in International Relations." New York: Cambridge University Press.

Adler, Emanuel and Michael N. Barnett. 1998*b*. *Security Communities*. Number 62 *in* "Cambridge Studies in International Relations." New York: Cambridge University Press.

Adler, Selig. 1957. *The Isolationist Impulse: Its Twentieth-Century Reaction*. Toronto, Ontario: Collier-Macmillan Canada, Ltd.

Albert, Mathias, Lars-Erik Cederman and Alexander Wendt, eds. 2010. *New Systems Theories of World Politics*. Basingstoke: Palgrave.

Albertini, Luigi. 1980. *The Origins of the War of 1914*. Westport, CT: Greenwood Press.

Albrecht-Carrié, René. 1958. *A Diplomatic History of Europe Since the Congress of Vienna*. New York: Harper & Brothers.

Americas Watch Committee (U.S.), ed. 1987. *Human Rights in Nicaragua: 1986*. New York: Americas Watch Committee.

Anderson, Benedict. 1987. *Imagined Communities: Reflections on the Origin and Spread of Nationalism*. 4th impression ed. London: Verso.

Andrew, Christopher and Vasili Mitrokhin. 2001. *The Sword and the Shield: The Mitrokhin Archive and the Secret History of the KGB*. New York: Basic Books.

Andrew, Christopher and Vasili Mitrokhin. 2005. *The World Was Going Our Way: The KGB and the Battle for the Third World*. New York: Basic Books.

Angell, Norman. 1910. *The Great Illusion: A Study of the Relation of Military*

Power in Nations to Their Economic and Social Advantage. New York: G. P. Putnam's Sons.

Arendt, Hannah. 2006. *Eichmann in Jerusalem: A Report on the Banality of Evil*. New York: Penguin Books.

Artz, Frederick B. 1934. *Reaction and Revolution, 1814–1832*. The Rise of Modern Europe (Series). New York: Harper & Row.

Axelrod, Robert. 1984. *The Evolution of Cooperation*. New York: Basic Books.

Bajoria, Jayshree. 2008. "The China-North Korea Relationship." *Washington Post*. **URL:** *http://www.washingtonpost.com/wp-dyn/content/article/2008/07/02/AR2008070201133.html*

Bara, C. 2014. "Incentives and Opportunities: A Complexity-Oriented Explanation of Violent Ethnic Conflict." *Journal of Peace Research* 51(6):696–710.

Barbieri, Katherine. 1996. "Economic Interdependence: A Path to Peace or a Source of Interstate Conflict?" *Journal of Peace Research* 33(1):29–49.

Bas, Muhammet and Robert Schub. 2016. "Mutual Optimism as a Cause of Conflict: Secret Alliances and Conflict Onset." *International Studies Quarterly* 60(3):552–564.

Bass, Gary Jonathan. 2008. *Freedom's Battle: The Origins of Humanitarian Intervention*. 1st ed. New York: Alfred A. Knopf.

Bearce, David H., Kristen M. Flanagan and Katharine M. Floros. 2006. "Alliances, Internal Information, and Military Conflict Among Member-States." *International Organization* 60(3):595–625.

Bearce, David H. and Sawa Omori. 2005. "How Do Commercial Institutions Promote Peace?" *Journal of Peace Research* 42(6):659–678.

Bell, David A. 2018. "Waiting for Steven Pinker's Enlightenment." **URL:** *https://www.thenation.com/article/waiting-for-steven-pinkers-enlightenment/*

Bennett, D. Scott and Allan C. Stam. 2003. *The Behavioral Origins of War*. Ann Arbor: University of Michigan Press.

Benson, Brett V. 2011. "Unpacking Alliances: Deterrent and Compellent Alliances and Their Relationship with Conflict, 1816–2000." *Journal of Politics* 73(4):1111–1127.

Berghahn, Volker R. 1973. *Germany and the Approach of War in 1914*. New York: St. Martin's.

Bhavnani, Ravi and Dan Miodownik. 2008. "Ethnic Polarization, Ethnic Salience, and Civil War." *Journal of Conflict Resolution* 53(1):30–49.

Bially Mattern, Janice and Ayşe Zarakol. 2016. "Hierarchies in World Politics." *International Organization* 70(3):623–654.

Binkley, Robert C. 1963. *Realism and Nationalism, 1852–1871*. The Rise of Modern Europe. New York: Harper and Row.

Blainey, Geoffrey. 1973. *The Causes of War*. New York: Free Press.

Blainey, Geoffrey. 2006. *A Short History of the 20th Century*. Chicago: I.R. Dee.

Boehmer, Charles, Erik Gartzke, and Timothy Nordstrom. 2004. "Do Intergovernmental Organizations Promote Peace?" *World Politics* 57(1): 1–38.

Boehmer, Charles R. and David Sobek. 2005. "Violent Adolescence: State Development and the Propensity for Militarized Interstate Conflict." *Journal of Peace Research* 42(1):5–26.

Bourne, Kenneth. 1970. *The Foreign Policy of Victorian England, 1830–1902.* Oxford: Clarendon Press.

Braumoeller, Bear F. 1997. "Deadly Doves: Liberal Nationalism and the Democratic Peace in the Soviet Successor States." *International Studies Quarterly* 41(3):375–402.

Braumoeller, Bear F. 1998. "Isolationism in International Relations." Ph.D. Thesis. University of Michigan.

Braumoeller, Bear F. 2010. "The Myth of American Isolationism." *Foreign Policy Analysis* 6(4):349–371.

Braumoeller, Bear F. and Austin Carson. 2011. "Political Irrelevance, Democracy, and the Limits of Militarized Conflict." *Journal of Conflict Resolution* 55(2):292–320.

Braumoeller, Bear F., Giampiero Marra, Rosalba Radice and Aisha E. Bradshaw. 2018. "Flexible Causal Inference for Political Science." *Political Analysis* 26(1):54–71.

Brecher, Michael and Jonathan Wilkenfeld. 1997. *A Study of Crisis.* Ann Arbor: University of Michigan Press.

Brecke, Peter. 1999. "Violent Conflicts 1400 A.D. to the Present in Different Regions of the World." Paper prepared for the 1999 Meeting of the Peace Science Society (International) on October 8–10, 1999, Ann Arbor, Michigan.

Brewer, David. 2001. *The Greek War of Independence: The Struggle for Freedom from Ottoman Oppression and the Birth of the Modern Greek Nation.* New York: Overlook Press.

Bridge, Roy. 1979. Allied Diplomacy in Peacetime: The Failure of the Congress System, 1815–23. In *Europe's Balance of Power, 1815–1848*, edited by Alan Sked, 34–53. New York: Barnes & Noble.

Bueno de Mesquita, Bruce, Alastair Smith, Randolph M. Siverson, and James D. Morrow. 2003. *The Logic of Political Survival.* Cambridge, MA: MIT Press.

Buhaug, Halvard. 2005. "Dangerous Dyads Revisited: Democracies May Not Be That Peaceful After All." *Conflict Management and Peace Science* 22(2):95–111.

Buhaug, Halvard. 2010. "Climate Not to Blame for African Civil Wars." *Proceedings of the National Academy of Sciences* 107(38):16477–16482.

Buhaug, Halvard, Nils Petter Gleditsch, and Ole Magnus Theisen. 2010. "Implications of Climate Change for Armed Conflict." In *Social Dimensions of Climate Change: Equity and Vulnerability in a Warming World*, edited by Andrew Norton, 75–101. Washington, DC: World Bank Publications.

Bull, Hedley. 1977. *The Anarchical Society: A Study of Order in World Politics.* New York: Columbia University Press.

Bulwer, Sir Henry Lytton. 1871. *The Life of Henry John Temple, Viscount Palmerston: With Selections from His Diaries and Correspondence.* Vol. 3. London: Richard Bentley.

Burke, Edmund. 1790. *Reflections on the Revolution in France, And on the Proceedings in Certain Societies in London Relative to That Event. In a Letter*

Intended to Have Been Sent to a Gentleman in Paris. 1st ed. London: J. Dodsley, in Pall-Mall.

Burkeman, Oliver. 2017. "Is the World Really Better Than Ever?" *The Guardian*. **URL:** *https://www.theguardian.com/news/2017/jul/28/is-the-world-really-better-than-ever-the-new-optimists*

Burleigh, Michael. 2001. *The Third Reich: A New History*. 1st American ed. New York: Hill and Wang.

Busby, Joshua W., Todd G. Smith, Kaiba L. White and Shawn M. Strange. 2013. "Climate Change and Insecurity: Mapping Vulnerability in Africa." *International Security* 37(4):132–172.

Bushman, Brad J., Patrick E. Jamieson, Ilana Weitz and Daniel Romer. 2013. "Gun Violence Trends in Movies." *Pediatrics* 132(6):1014–1018.

Bussmann, M. 2010. "Foreign Direct Investment and Militarized International Conflict." *Journal of Peace Research* 47(2):143–153.

Butler, Christopher K. and Scott Gates. 2012. "African Range Wars: Climate, Conflict, and Property Rights." *Journal of Peace Research* 49(1):23–34.

Butterfield, Herbert. 1965. *The Whig Interpretation of History*. Number 318 *in* "The Norton Library." New York: Norton.

Buzan, Barry. 2017. Universal Sovereignty. In *The Globalization of International Society*, edited by Tim Dunne and Christian Reus-Smit, 227–247. New York: Oxford University Press.

Calleo, David. 1978. *The German Problem Reconsidered: Germany and the World Order, 1870 to the Present*. London: Cambridge University Press.

Campbell, Susanna P., Michael G. Findley and Kyosuke Kikuta. 2017. "An Ontology of Peace: Landscapes of Conflict and Cooperation with Application to Colombia." *International Studies Review* 19(1):92–113.

Caporaso, James A. 2000. "Changes in the Westphalian Order: Territory, Public Authority, and Sovereignty." *International Studies Review* 2(2):1–28.

Caprioli, Mary and Peter F. Trumbore. 2005. "Rhetoric Versus Reality: Rogue States in Interstate Conflict." *Journal of Conflict Resolution* 49(5):770–791.

Carr, Edward Hallett. 1939. *The Twenty Years' Crisis, 1919–1939: An Introduction to the Study of International Relations*. New York: Harper and Row.

Carswell, A. J. 2013. "Unblocking the UN Security Council: The Uniting for Peace Resolution." *Journal of Conflict and Security Law* 18(3):453–480.

Cederman, Lars-Erik. 1997. *Emergent Actors in World Politics*. Princeton, NJ: Princeton University Press.

Cederman, Lars-Erik. 2001. "Back to Kant: Reinterpreting the Democratic Peace as a Macrohistorical Learning Process." *American Political Science Review* 95(1):15–31.

Cederman, Lars-Erik. 2003. "Modeling the Size of Wars: From Billiard Balls to Sandpiles." *American Political Science Review* 97(1):135–150.

Cederman, Lars-Erik, Andreas Wimmer and Brian Min. 2009. "Why Do Ethnic Groups Rebel? New Data and Analysis." *World Politics* 62(1): 87–119.

Cederman, Lars-Erik, Halvard Buhaug and Jan Ketil Rod. 2009. "Ethno-Nationalist Dyads and Civil War: A GIS-Based Analysis." *Journal of Conflict Resolution* 53(4):496–525.

Cederman, Lars-Erik, Kristian Skrede Gleditsch, Idean Salehyan and Julian Wuchterpfennig. 2013. "Transborder Ethnic Kin and Civil War." *International Organization* 67(2):389–410.

Cederman, Lars-Erik and Luc Girardin. 2007. "Beyond Fractionalization: Mapping Ethnicity Onto Nationalist Insurgencies." *American Political Science Review* 101(1):173–185.

Cederman, Lars-Erik, T. Warren and Didier Sornette. 2011. "Testing Clausewitz: Nationalism, Mass Mobilization, and the Severity of War." *International Organization* 65(4):605–638.

Chapman, Thomas and Philip G. Roeder. 2007. "Partition as a Solution to Wars of Nationalism: The Importance of Institutions." *American Political Science Review* 101(4):677–691.

Chaudoin, Stephen, Helen V. Milner and Xun Pang. 2015. "International Systems and Domestic Politics: Linking Complex Interactions With Empirical Models in International Relations." *International Organization* 69(2): 275–309.

Chiba, Daina, Carla Martinez Machain and William Reed. 2014. "Major Powers and Militarized Conflict." *Journal of Conflict Resolution* 58(6):976–1002.

Choi, S.-W. 2010*a*. "Legislative Constraints: A Path to Peace?" *Journal of Conflict Resolution* 54(3):438–470.

Choi, Seung-Whan. 2010*b*. "Beyond Kantian Liberalism: Peace Through Globalization?" *Conflict Management and Peace Science* 27(3):272–295.

Christensen, Thomas J. 1997. *Useful Adversaries: Grand Strategy, Domestic Mobilization, and Sino-American Conflict, 1947–1958*. Princeton, NJ: Princeton University Press.

Christensen, Thomas and Jack Snyder. 1990. "Chain Gangs and Passed Bucks: Predicting Alliance Patterns in Multipolarity." *International Organization* 44(2):137–168.

Cirillo, Pasquale and Nassim Nicholas Taleb. 2015. "On the Tail Risk of Violent Conflict and Its Underestimation." *eprint arXiv:1505.04722*. **URL:** *http://arxiv.org/abs/1505.04722v1*

Cirillo, Pasquale and Nassim Nicholas Taleb. 2016*a*. The Decline of Violent Conflicts: What Do the Data Really Say? In *Nobel Symposium 161: The Causes of Peace*. Bergen, Norway.

Cirillo, Pasquale and Nassim Nicholas Taleb. 2016*b*. "On the Statistical Properties and Tail Risk of Violent Conflicts." *Physica A: Statistical Mechanics and Its Applications* 452:29–45.

Clark, Christopher M. 2013. *The Sleepwalkers: How Europe Went to War in 1914*. First U.S. ed. New York: Harper.

Clark, Ian. 1989. *The Hierarchy of States: Reform and Resistance in the International Order*. Number 7 *in* "Cambridge Studies in International Relations." Cambridge: Cambridge University Press.

Claude, Inis L. 1966. "Collective Legitimization as a Political Function of the United Nations." *International Organization* 20(3):367–379.

Clauset, Aaron. 2018. "Trends and Fluctuations in the Severity of Interstate Wars." *Science Advances* 4(2):eaao3580.

Clauset, Aaron, Cosma Rohilla Shalizi and M. E. J. Newman. 2009. "Power-Law Distributions in Empirical Data." *SIAM Review* 51(4):661–703.

Cobden, Richard. 1997. Commerce Is the Grand Panacea. In *The Libertarian Reader*, edited by David Boaz, 320–231. New York: Free Press.

Cohen, Youssef, Brian R. Brown and A. F. K. Organski. 1981. "The Paradoxical Nature of State Making: The Violent Creation of Order." *American Political Science Review* 75(4):901–910.

Cook, Sherburne F. 1956. "The Aboriginal Population of the North Coast of California." *Anthropological Records* 16:81–130.

Craig, Gordon A. 1965. *From Bismarck to Adenauer: Aspects of German Statecraft*. New York: Harper and Row.

Craig, Gordon A. 1978. *Germany, 1866–1945*. New York: Oxford University Press.

Craig, Gordon A. and Alexander L. George. 1995. *Force and Statecraft: Diplomatic Problems of Our Time*. 3rd ed. New York: Oxford University Press.

Crankshaw, Edward. 1981. *Bismarck*. New York: Viking Press.

Crescenzi, M. J.C., J. D. Kathman and S. B. Long. 2007. "Reputation, History, and War." *Journal of Peace Research* 44(6):651–667.

Croco, Sarah E. and Scott Sigmund Gartner. 2014. "Flip-Flops and High Heels: An Experimental Analysis of Elite Position Change and Gender on Wartime Public Support." *International Interactions* 40(1):1–24.

Cronin, Bruce. 2010. "Be Careful What You Wish For: War Aims and the Construction of Postwar Political Orders." *Journal of Peace Research* 47(6):791–801.

Cudworth, Erika and Stephen Hobden. 2010. "Anarchy and Anarchism: Towards a Theory of Complex International Systems." *Millennium: Journal of International Studies* 39(2):399–416.

Cunen, Céline, Nils Lid Hjort and Håvard Mokleiv Nygård. 2018. "Statistical Sightings of Better Angels: Analysing the Distribution of Battle Deaths in Interstate Conflict Over Time." Manuscript, University of Oslo.

Cunningham, Kathleen Gallagher. 2013. "Actor Fragmentation and Civil War Bargaining: How Internal Divisions Generate Civil Conflict." *American Journal of Political Science* 57(3):659–672.

Daalder, Ivo H. and James G. Stavridis. 2012. "NATO's Victory in Libya: The Right Way to Run an Intervention." *Foreign Affairs* 91(2):2–7.

Dafoe, Allan, John R. Oneal, and Bruce Russett. 2013. "The Democratic Peace: Weighing the Evidence and Cautious Inference." *International Studies Quarterly* 57(1):201–214.

Davis, David R. and Will H. Moore. 1997. "Ethnicity Matters: Transnational Ethnic Alliances and Foreign Policy Behavior." *International Studies Quarterly* 41(1):171–184.

Dawson, William Harbutt. 1919. *The German Empire, 1867–1914, and the Unity Movement*. Vol. 2. New York: Macmillan.

Dean, Richard N. 1980. "Beyond Helsinki: The Soviet View of Human Rights in International Law." *Virginia Journal of International Law* 21:55–95.

Dedijer, Vladimir. 1966. *The Road to Sarajevo*. New York: Simon and Schuster.

DeNardo, James. 1997. Complexity, Formal Methods, and Ideology in International Studies. In *New Thinking in International Relations Theory*, edited by Michael W. Doyle and G. John Ikenberry, 124–162. Boulder, CO: Westview.

Deutsch, Karl W., Sidney A. Burrell, Robert A. Kann, Maurice Lee, Jr., Martin Lichterman, Raymond Lindgren, Francis L. Loewenheim, and Richard W. Van Wagenen. 1957. *Political Community and the North Atlantic Area: International Organization in the Light of Historical Experience*. Princeton, NJ: Princeton University Press.

Diehl, Paul F. 1983. "Arms Races and Escalation: A Closer Look." *Journal of Peace Research* 20(3):205–212.

Diehl, Paul F. 1992. "What Are They Fighting For? The Importance of Issues in International Conflict Research." *Journal of Peace Research* 29(3):333–344.

Diehl, Paul F. and Mark J. C. Crescenzi. 1998. "Reconfiguring the Arms Race–War Debate." *Journal of Peace Research* 35(1):111–118.

Diehl, Paul F., Jennifer Reifschneider, and Paul R. Hensel. 1996. "United Nations Intervention and Recurring Conflict." *International Organization* 50(4):683–700.

Dillon, Sam. 1992. *Comandos: The CIA and Nicaragua's Contra Rebels*. New York: H. Holt.

Dorussen, H. 2006. "Heterogeneous Trade Interests and Conflict: What You Trade Matters." *Journal of Conflict Resolution* 50(1):87–107.

Dorussen, Han and Hugh Ward. 2008. "Intergovernmental Organizations and the Kantian Peace: A Network Perspective." *Journal of Conflict Resolution* 52(2):189–212.

Dorussen, Han and Hugh Ward. 2010. "Trade Networks and the Kantian Peace." *Journal of Peace Research* 47(1):29–42.

Dower, John W. 1986. *War Without Mercy: Race and Power in the Pacific War*. 1st ed. New York: Pantheon Books.

Doyle, Michael W. 1983*a*. "Kant, Liberal Legacies, and Foreign Affairs." *Philosophy and Public Affairs* 12(3):205–235.

Doyle, Michael W. 1983*b*. "Kant, Liberal Legacies, and Foreign Affairs, Part 2." *Philosophy and Public Affairs* 12(4):323–353.

Duras, Victor Hugo. 1908. *Universal Peace*. New York: Broadway.

Eckhardt, William. 1989. "Civilian Deaths in Wartime." *Bulletin of Peace Proposals* 20(1):89–98.

Edsell, Thomas B. 2017. "Trump Knows How to Push Our Buttons." *New York Times*. **URL:** *https://www.nytimes.com/2017/08/03/opinion/donald-trump-democrats-better-deal.html*

Elias, Norbert. 1988. Violence and Civilization: The State Monopoly of Physical Violence and Its Infringement. In *Civil Society and the State: New European Perspectives*, edited by John Keane, 177–198. New York: Verso.

Elias, Norbert. 2000 (1939). *The Civilizing Process: Sociogenetic and Psychogenetic Investigations*. Rev. ed. Malden, MA: Blackwell Publishers.

Elrod, Richard B. 1976. "The Concert of Europe: A Fresh Look at an International System." *World Politics* 28(2):159–174.

Ensor, Robert Charles Kirkwood. 1936. *England, 1870–1914*. The Oxford History of England. Oxford: Clarendon Press.

Epstein, Joshua M. 1999. "Agent-Based Computational Models and Generative Social Science." *Complexity* 4(5):41–60.

Eyck, Erich. 1964. *Bismarck and the German Empire*. New York: W. W. Norton and Co.

Falk, Dean and Charles Hildebolt. 2017. "Annual War Deaths in Small-Scale versus State Societies Scale with Population Size Rather than Violence." *Current Anthropology* 58(6):805–813.

Fang, Songying, Jesse C. Johnson and Brett Ashley Leeds. 2014. "To Concede or to Resist? The Restraining Effect of Military Alliances." *International Organization* 68(4):775–809.

Fazal, Tanisha M. 2014. "Dead Wrong? Battle Deaths, Military Medicine, and Exaggerated Reports of War's Demise." *International Security* 39(1):95–125.

Fearon, James D. 1995. "Rationalist Explanations for War." *International Organization* 49(3):379–414.

Fearon, James D. 1998. "Bargaining, Enforcement, and International Cooperation." *International Organization* 52(2):269–305.

Fearon, James D. and David Laitin. 2003. "Ethnicity, Insurgency, and Civil War." *American Political Science Review* 97(1):75–90.

Ferguson, Niall. 2008. *Virtual History: Alternatives and Counterfactuals*. New York: Basic Books.

Ferguson, R. Brian. 2013. "Pinker's List: Exaggerating Prehistoric War Mortality." In *War, Peace, and Human Nature: The Convergence of Evolutionary and Cultural Views*, edited by Douglas P. Fry, 112–131. New York: Oxford University Press.

Finer, Samuel E. 1997. *The History of Government from the Earliest Times*. Oxford: Oxford University Press.

Firchow, Pamina and Roger Mac Ginty. 2017. "Measuring Peace: Comparability, Commensurability, and Complementarity Using Bottom-Up Indicators." *International Studies Review* 19(1):6–27.

Fischer, Fritz. 1975. *War of Illusions: German Policies From 1911 to 1914*. New York: W. W. Norton.

Fortna, Page. 2013. "Has Violence Declined in World Politics?" *Perspectives on Politics* 11(2):566–570.

Fortna, Virginia Page. 2004*a*. "Interstate Peacekeeping: Causal Mechanisms and Empirical Effects." *World Politics* 56(4):481–519.

Fortna, Virginia Page. 2004*b*. *Peace Time: Cease-Fire Agreements and the Durability of Peace*. Princeton, NJ: Princeton University Press.

Fravel, M. Taylor. 2008. *Strong Borders, Secure Nation: Cooperation and Conflict in China's Territorial Disputes*. Princeton, NJ: Princeton University Press.

Freedman, Lawrence. 2017. *The Future of War: A History*. First ed. New York: Public Affairs.

Friedman, Jeremy Scott. 2015. *Shadow Cold War: The Sino-Soviet Competition for the Third World.* Chapel Hill: University of North Carolina Press.
Fukuyama, Francis. 2011. *The Origins of Political Order: From Prehuman Times to the French Revolution.* 1st ed. New York: Farrar, Straus and Giroux.
Fukuyama, Francis. 2014. *Political Order and Political Decay: From the Industrial Revolution to the Globalization of Democracy.* 1st ed. New York: Farrar, Straus and Giroux.
Gaddis, John Lewis. 1986. "The Long Peace: Elements of Stability in the Postwar International System." *International Security* 10(4):99–142.
Gaddis, John Lewis. 1997. *We Now Know.* New York: Oxford University Press.
Gaddis, John Lewis. 2005. *The Cold War: A New History.* New York: Penguin Press.
Galtung, Johan. 1967. "Theories of Peace: A Synthetic Approach to Peace Thinking." Manuscript, International Peace Research Institute, Oslo, Norway. **URL:** *https://www.transcend.org/files/Galtung_Book_unpub_Theories_of_Peace_-_A_Synthetic_Approach_to_Peace_Thinking_1967.pdf*
Gartzke, Erik, Quan Li and Charles Boehmer. 2001. "Investing in the Peace: Economic Interdependence and International Conflict." *International Organization* 55(2):391–438.
Gat, Azar. 2008. *War in Human Civilization.* Oxford: Oxford University Press.
Gat, Azar. 2013. "Is War Declining—and Why?" *Journal of Peace Research* 50(2):149–157.
Geller, Daniel S. and J. David Singer. 1998. *Nations at War: A Scientific Study of International Conflict.* Cambridge: Cambridge University Press.
Gelpi, Christopher. 1999. Alliances as Instruments of Intra-Allied Control. In *Imperfect Unions: Security Institutions Over Time and Space*, edited by Helga Haftendorn, Celeste Wallander and Robert Keohane, 107–139. New York: Oxford University Press.
Ghobarah, Hazem Adam, Paul Huth, and Bruce Russett. 2003. "Civil Wars Kill and Maim People—Long After the Shooting Stops." *American Political Science Review* 97(2):189–202.
Gibler, Douglas M. 2007. "Bordering on Peace: Democracy, Territorial Issues, and Conflict." *International Studies Quarterly* 51(3):509–532.
Gibler, Douglas M., Steven V. Miller, and Erin K. Little. 2016. "An Analysis of the Militarized Interstate Dispute (MID) Dataset, 1816–2001." *International Studies Quarterly* 60(4):719–730.
Gilpin, Robert. 1981. *War and Change in World Politics.* Cambridge: Cambridge University Press.
Gleditsch, Kristian. 2004. "A Revised List of Wars Between and Within Independent States, 1816–2002." *International Interactions* 30(3):231–262.
Gleditsch, Kristian S. and Michael D. Ward. 1999. "A Revised List of Independent States Since the Congress of Vienna." *International Interactions* 25(4):393–413.

Gleditsch, N. P., P. Wallensteen, M. Eriksson, M. Sollenberg and H. Strand. 2002. "Armed Conflict 1946–2001: A New Dataset." *Journal of Peace Research* 39(5):615–637.

Goddard, Stacie E. 2006. "Uncommon Ground: Indivisible Territory and the Politics of Legitimacy." *International Organization* 60(1):35–68.

Goenner, C. F. 2010. "From Toys to Warships: Interdependence and the Effects of Disaggregated Trade on Militarized Disputes." *Journal of Peace Research* 47(5):547–559.

Goertz, Gary, Paul F. Diehl, and Alexandru Balas. 2016. *The Puzzle of Peace: The Evolution of Peace in the International System.* New York: Oxford University Press.

Goldsmith, B. E., S. K. Chalup and M. J. Quinlan. 2008. "Regime Type and International Conflict: Towards a General Model." *Journal of Peace Research* 45(6):743–763.

Goldstein, Joshua S. 1988. *Long Cycles: Prosperity and War in the Modern Age.* New Haven, CT: Yale University Press.

Goldstein, Joshua S. 2011. *Winning the War on War: The Decline of Armed Conflict Worldwide.* New York: Dutton Adult.

Goldstein, Joshua S. and Steven Pinker. 2016. "The Decline of War and Violence." *Boston Globe.* **URL:** *https://www.bostonglobe.com/opinion/2016/04/15/the-decline-war-and-violence/lxhtEplvpptoBz9kPphzkL/story.html*

Gorodetsky, Gabriel. 1994. The Formulation of Soviet Foreign Policy—Ideology and Realpolitik. In *Soviet Foreign Policy, 1917–1991*, edited by Gabriel Gorodetsky, 30–44. London: Frank Cass and Co. Ltd.

Grane, William Leighton. 1912. *The Passing of War: A Study in Things That Make for Peace.* London: Macmillan.

Gray, John. 2015. "John Gray: Steven Pinker Is Wrong about Violence and War." *The Guardian.* **URL:** *https://www.theguardian.com/books/2015/mar/13/john-gray-steven-pinker-wrong-violence-war-declining*

Gulick, Edward Vose. 1955. *Europe's Classical Balance of Power.* New York: W. W. Norton.

Gwertzman, Bernard. 1982. "Reagan, in a Phone Call, Tried to Deter Invasion." *New York Times.* **URL:** *http://www.nytimes.com/1982/04/03/world/reagan-in-a-phone-call-tried-to-deter-invasion.html*

Hafner-Burton, Emilie M. and Alexander H. Montgomery. 2006. "Power Positions: International Organizations, Social Networks, and Conflict." *Journal of Conflict Resolution* 50(1):3–27.

Hafner-Burton, Emilie M. and Alexander H. Montgomery. 2012. "War, Trade, and Distrust: Why Trade Agreements Don't Always Keep the Peace." *Conflict Management and Peace Science* 29(3):257–278.

Halberstam, David. 2001. *The Best and the Brightest.* New York: Modern Library.

Hall, John A. and Siniša Malešević, eds. 2013. *Nationalism and War*. Cambridge: Cambridge University Press.

Hanlon, Aaron R. 2018. "Steven Pinker's New Book on the Enlightenment Is a Huge Hit. Too Bad It Gets the Enlightenment Wrong." **URL:** *https://www.vox.com/the-big-idea/2018/5/17/17362548/pinker-enlightenment-now-two-cultures-rationality-war-debate*

Hartmann, Anja and Beatrice Heuser, eds. 2001. *War, Peace, and World Orders in European History*. London and New York: Routledge.

Hastie, Trevor and Robert Tibshirani. 1990. *Generalized Additive Models*. Boca Raton, FL: Chapman & Hall/CRC.

Hastings, Max and Simon Jenkins. 1983. *The Battle for the Falklands*. New York: Norton.

Hayes, Carlton J. H. 1931. *The Historical Evolution of Modern Nationalism*. New York: Richard R. Smith, Inc.

Haynes, K. 2012. "Lame Ducks and Coercive Diplomacy: Do Executive Term Limits Reduce the Effectiveness of Democratic Threats?" *Journal of Conflict Resolution* 56(5):771–798.

Healy, Brian and Arthur Stein. 1973. "The Balance of Power in International History." *Journal of Conflict Resolution* 17(1):33–61.

Hegre, Håvard. 2004. "The Duration and Termination of Civil War." *Journal of Peace Research* 41(3):243–252.

Hegre, Håvard. 2008. "Gravitating Toward War: Preponderance May Pacify, but Power Kills." *Journal of Conflict Resolution* 52(4):566–589.

Hendrix, Cullen S. and Sarah M. Glaser. 2007. "Trends and Triggers: Climate, Climate Change, and Civil Conflict in Sub-Saharan Africa." *Political Geography* 26(6):695–715.

Hendrix, Cullen S. and Idean Salehyan. 2012. "Climate Change, Rainfall, and Social Conflict in Africa." *Journal of Peace Research* 49(1):35–50.

Henrich, Joseph, Steven J. Heine, and Ara Norenzayan. 2010. "The Weirdest People in the World?" *Behavioral and Brain Sciences* 33(2-3):61-83.

Hensel, Paul R. 2001. "Contentious Issues and World Politics: The Management of Territorial Claims in the Americas, 1816–1992." *International Studies Quarterly* 45(1):81–109.

Hensel, Paul R. and Sara McLaughlin Mitchell. 2005. "Issue Indivisibility and Territorial Claims." *GeoJournal* 64(4):275–285.

Herrmann, David G. 1996. *The Arming of Europe and the Making of the First World War*. Princeton, NJ: Princeton University Press.

Herz, John H. 1951. *Political Realism and Political Idealism: A Study in Theories and Realities*. Chicago: University of Chicago Press.

Hinsley, F. H. 1963. *Power and the Pursuit of Peace: Theory and Practice in the History of Relations Between States*. Cambridge: Cambridge University Press.

Hitchens, Christopher. 2005. "Ohio's Odd Numbers." **URL:** *https://www.vanityfair.com/news/2005/03/hitchens200503*

Hobbes, Thomas. 1997 (1651). *Leviathan*. New York: W. W. Norton.

Holborn, Hajo. 1969. *A History of Modern Germany, 1840–1945*. Princeton, NJ: Princeton University Press.

Holland, John H. 2014. *Complexity: A Very Short Introduction*. Oxford: Oxford University Press.

Holm, Sture. 1979. "A Simple Sequentially Rejective Multiple Test Procedure." *Scandinavian Journal of Statistics* 6(2):65–70.

Holsti, Kalevi J. 1991. *Peace and War: Armed Conflicts and International Order, 1648–1989*. Cambridge: Cambridge University Press.

Holsti, Ole R. 1979. "The Three-Headed Eagle: The United States and System Change." *International Studies Quarterly* 23(3):339–359.

Hopf, Ted. 1991. "Polarity, the Offense-Defense Balance, and War." *American Political Science Review* 85(2):475–493.

Horowitz, M. 2009. "The Spread of Nuclear Weapons and International Conflict: Does Experience Matter?" *Journal of Conflict Resolution* 53(2):234–257.

Horowitz, Michael C. and Idean Salehyan. 2015. "Joe Public v. Sue Scholar: Support for the Use of Force." **URL:** *http://politicalviolenceataglance.org/2015/07/27/joe-public-v-sue-scholar-support-for-the-use-of-force/*

Howard, Lise Morjé and Alexandra Stark. 2018. "How Civil Wars End: The International System, Norms, and the Role of External Actors." *International Security* 42(3):127–171.

Howard, Michael. 1991. *The Lessons of History*. New Haven, CT: Yale University Press.

Howard, Michael. 2000. *The Invention of Peace: Reflections on War and International Order*. New Haven, CT: Yale University Press.

Huntington, Samuel P. 1958. "Arms Races: Prerequisites and Results." *Public Policy* 8(1):41–83.

Huntington, Samuel P. 1968. *Political Order in Changing Societies*. New Haven, CT: Yale University Press.

Hurd, Ian. 1999. "Legitimacy and Authority in International Politics." *International Organization* 53(2):379–408.

Huth, Paul K. and Todd L. Allee. 2002. *The Democratic Peace and Territorial Conflict in the Twentieth Century*. Cambridge: Cambridge University Press.

Hwang, Wonjae. 2010. "Power, Preferences, and Multiple Levels of Interstate Conflict." *International Interactions* 36(3):215–239.

Ikenberry, G. John. 2001. *After Victory: Institutions, Strategic Restraint, and the Rebuilding of Order After Major Wars*. Princeton, NJ: Princeton University Press.

Ikenberry, G. John. 2011. *Liberal Leviathan: The Origins, Crisis, and Transformation of the American World Order*. Princeton Studies in International History and Politics. Princeton, NJ: Princeton University Press.

Ikenberry, G. John. 2014*a*. Introduction: Power, Order, and Change in World Politics. In *Power, Order, and Change in World Politics*, edited by G. John Ikenberry, 1–16. New York: Cambridge University Press.

Ikenberry, G. John. 2014*b*. The Logic of Order: Westphalia, Liberalism, and the

Evolution of International Order in the Modern Era. In *Power, Order, and Change in World Politics*, edited by G. John Ikenberry, 83–106. New York: Cambridge University Press.

International Coalition for the Responsibility to Protect. 2017. "RtoP and Rebuilding: The Role of the Peacebuilding Commission." **URL:** *http://www.responsibilitytoprotect.org/index.php/about-rtop/related-themes/2417-pbc-and-rtop*

Iriye, Akira. 1987. *The Origins of the Second World War in Asia and the Pacific*. New York: Longman.

Jäckel, Eberhard. 1972. *Hitler's Weltanschauung: Blueprint for Power*. Middletown, CT: Wesleyan University Press.

James, Nicholas A. and David S. Matteson. 2014. "ecp: An R Package for Nonparametric Multiple Change Point Analysis of Multivariate Data." *Journal of Statistical Software* 62(7):1–25.

Janik, Ralph R. A. 2017. The Responsibility to Protect, Conflict Prevention, and the *Ius Ad Bellum*: What Role for Democracy? In *Gerechte Intervention? Zwischen Gewaltverbot Und Schutzverantwortung*, edited by Stephanie Fenkart, Heinz Gärtner and Hannes Swoboda, 167–190. LIT Verlag.

Jarausch, Konrad H. 1969. "The Illusion of Limited War: Chancellor Bethmann Hollweg's Calculated Risk, July 1914." *Central European History* 2(1):48–76.

Jarrett, Mark. 2013. *The Congress of Vienna and Its Legacy: War and Great Power Diplomacy After Napoleon*. London: I.B. Tauris.

Jenke, Libby and Christopher Gelpi. 2017. "Theme and Variations: Historical Contingencies in the Causal Model of Interstate Conflict." *Journal of Conflict Resolution* 61(10):2262–2284.

Jenkins, Roy. 1997. *Gladstone: A Biography*. New York: Random House.

Jervis, Robert. 1976. *Perception and Misperception in International Politics*. Princeton, NJ: Princeton University Press.

Jervis, Robert. 1978. "Cooperation Under the Security Dilemma." *World Politics* 30(2):167–214.

Jervis, Robert. 1982. "Security Regimes." *International Organization* 36(2):357–378.

Jervis, Robert. 1985. "From Balance to Concert: A Study of International Security Cooperation." *World Politics* 38(1):58–79.

Jervis, Robert. 1993. "Arms Control, Stability, and Causes of War." *Political Science Quarterly* 108(2):239–253.

Jervis, Robert. 1997. *System Effects: Complexity in Political and Social Life*. Princeton, NJ: Princeton University Press.

Jervis, Robert. 2011. "Pinker the Prophet." *National Interest* 116(Nov–Dec):54–64.

Johnson, Dominic D.P. and Monica Duffy Toft. 2014. "Grounds for War: The Evolution of Territorial Conflict." *International Security* 38(3):7–38.

Johnson, Jesse C. and Tiffany D. Barnes. 2011. "Responsibility and the Diversionary Use of Force 1." *Conflict Management and Peace Science* 28(5):478–496.

Joll, James. 1992. *The Origins of the First World War*. Origins of Modern Wars. 2nd ed. London and New York: Longman.

Jonas, Manfred. 1966. *Isolationism in America, 1935–1941*. Ithaca, NY: Cornell University Press.

Jones, Benjamin T., Eleonora Mattiacci and Bear F. Braumoeller. 2017. "Food Scarcity and State Vulnerability: Unpacking the Link Between Climate Variability and Violent Unrest." *Journal of Peace Research* 54(3):335–350.

Jones, D. M., S. A. Bremer and J. D. Singer. 1996. "Militarized Interstate Disputes, 1816–1992: Rationale, Coding Rules, and Empirical Patterns." *Conflict Management and Peace Science* 15(2):163–213.

Judt, Tony. 2005. *Postwar: A History of Europe Since 1945*. New York: Penguin Press.

Jung, Sung Chul. 2014. "Foreign Targets and Diversionary Conflict." *International Studies Quarterly* 58(3):566–78.

Kagan, Donald. 1969. *The Outbreak of the Peloponnesian War*. Ithaca, NY: Cornell University Press.

Kagan, Robert. 1996. *A Twilight Struggle: American Power and Nicaragua, 1977–1990*. New York: Free Press.

Kalyvas, Stathis N. and Laia Balcells. 2010. "International System and Technologies of Rebellion: How the End of the Cold War Shaped Internal Conflict." *American Political Science Review* 104(3):415–429.

Kant, Immanuel. 2003. *To Perpetual Peace: A Philosophical Sketch*. Indianapolis, IN: Hackett Pub.

Kasten, Lukas. 2017. "When Less Is More: Constructing a Parsimonious Concept of Interstate Peace for Quantitative Analysis." *International Studies Review* 19(1):28–52.

Kaufmann, Chaim. 1996. "Possible and Impossible Solutions to Ethnic Civil Wars." *International Security* 20(4):136–175.

Keegan, John. 1993. *A History of Warfare*. New York: Vintage Books.

Keegan, John. 1999. *The First World War*. 1st American ed. New York: A. Knopf.

Kennedy, Paul. 1987. *The Rise and Fall of the Great Powers: Economic Change and Military Conflict From 1500 to 2000*. New York: Random House.

Kennedy, Paul M. 1980. *The Rise of the Anglo-German Antagonism, 1860–1914*. Boston: Allen & Unwin.

Keohane, Robert O. 1984. *After Hegemony: Cooperation and Discord in the World Political Economy*. Princeton, NJ: Princeton University Press.

Keohane, Robert O. and Lisa L. Martin. 1995. "The Promise of Institutionalist Theory." *International Security* 20(1):39–51.

Kertzer, Joshua D. 2013. "Making Sense of Isolationism: Foreign Policy Mood as a Multilevel Phenomenon." *Journal of Politics* 75(1):225–240.

Kim, Sung-han. 2003. "The End of Humanitarian Intervention?" *Orbis* 47(4):721–736.

Kissinger, Henry. 1957. *A World Restored*. London: Weidenfeld and Nicolson.

Kissinger, Henry. 1994. *Diplomacy*. New York: Simon and Schuster.

Kissinger, Henry. 2014. *World Order*. New York: Penguin Press.

Knock, Thomas J. 1992. *To End All Wars: Woodrow Wilson and the Quest for a New World Order*. Princeton, NJ: Princeton University Press.
Kohn, Hans. 1944. *The Idea of Nationalism: A Study in Its Origins and Background*. New York: Macmillan.
Kokoshin, Andrei A. 1998. *Soviet Strategic Thought, 1917–91*. Cambridge, MA: MIT Press.
Kordunsky, Anna. 2013. "Study Challenges Theory Modern Nations Are Less Warlike." **URL:** *http://news.nationalgeographic.com/news/study-challenges-theory-modern-nations-are-less-warlike/*
Krasner, Stephen D. 1982. "Structural Causes and Regime Consequences: Regimes as Intervening Variables." *International Organization* 36(2):185–205.
Kupchan, Charles. 2010. *How Enemies Become Friends: The Sources of Stable Peace*. Princeton, NJ: Princeton University Press.
Kupchan, Charles A. 2014. "Unpacking Hegemony: The Social Foundations of Hierarchical Order." In *Power, Order, and Change in World Politics*, edited by G. John Ikenberry, 19–60. New York: Cambridge University Press.
Kuper, Simon. 2014. "The Surprising Power of Peace." **URL:** *https://www.ft.com/content/b5df33b8-b539-11e3-af92-00144feabdc0*
Kydd, Andrew. 1997. "Game Theory and the Spiral Model." *World Politics* 49(3):371–400.
Kydd, Andrew. 2000. "Arms Races and Arms Control: Modeling the Hawk Perspective." *American Journal of Political Science* 44(2):222–238.
Laitin, David D. 2004. "Ethnic Unmixing and Civil War." *Security Studies* 13(4):350–365.
Lake, David A. 1992. "Powerful Pacifists: Democratic States and War." *American Political Science Review* 86(1):24–37.
Lake, David A. 2009. *Hierarchy in International Relations*. Ithaca, NY: Cornell University Press.
Lake, David A. 2014. "Dominance and Subordination in World Politics: Authority, Liberalism, and Stability in the Modern International Order." In *Power, Order, and Change in World Politics*, edited by G. John Ikenberry, 61–82. New York: Cambridge University Press.
Lake, David A. and Patrick M. Morgan, eds. 1997. *Regional Orders: Building Security in a New World*. University Park, PA: Pennsylvania State University Press.
Landini, Tatiana Savoia and François Dépelteau, eds. 2017. *Norbert Elias and Violence*. New York: Palgrave Macmillan.
Langer, William L. 1966. *European Alliances and Alignments, 1871–1890*. 2nd ed. New York: Alfred A. Knopf.
Langer, William L. and S. Everett Gleason. 1952. *The Challenge to Isolation: The World Crisis of 1937–1940 and American Foreign Policy*. New York: Harper and Brothers.

Latané, Bibb and Judith Rodin. 1969. "A Lady in Distress: Inhibiting Effects of Friends and Strangers on Bystander Intervention." *Journal of Experimental Social Psychology* 5(2):189–202.

Lebow, Richard Ned. 1984. "Windows of Opportunity: Do States Jump Through Them?" *International Security* 9(1):147–186.

Lebow, Richard Ned. 2010. *Forbidden Fruit: Counterfactuals and International Relations*. Princeton, NJ: Princeton University Press.

Leeds, Brett Ashley. 2003. "Do Alliances Deter Aggression? The Influence of Military Alliances on the Initiation of Militarized Interstate Disputes." *American Journal of Political Science* 47(3):427–439.

Lemann, Nicholas. 1996. "Kicking in Groups." **URL:** *https://www.theatlantic.com/magazine/archive/1996/04/kicking-in-groups/376562/*

Lenin, V. I. 1939. *Imperialism: The Highest Stage of Capitalism*. New York: International Publishers.

Levy, Jack S. 1983. *War in the Modern Great Power System, 1495–1975*. Lexington: University Press of Kentucky.

Levy, Jack S. 2013. "Has Violence Declined in World Politics?" *Perspectives on Politics* 11(2):573–577.

Levy, Jack S. and William R. Thompson. 2011. *The Arc of War: Origins, Escalation, and Transformation*. Chicago: University of Chicago Press.

Levy, Jack S. and William R. Thompson. 2013. "The Decline of War? Multiple Trajectories and Diverging Trends." *International Studies Review* 15(3):411–419.

Lewis, Patricia, Heather Williams, Benoît Pelopidas and Sasan Aghlani. 2014. "Too Close for Comfort: Cases of Near Nuclear Use and Options for Policy." **URL:** *https://www.chathamhouse.org/sites/default/files/field/field_document/20140428TooCloseforComfortNuclearUseLewisWilliamsPelopidasAghlani.pdf*

Lilla, Mark. 2016. *The Shipwrecked Mind: On Political Reaction*. New York: New York Review of Books.

Lipset, Seymour Martin. 1959. "Some Social Requisites of Democracy: Economic Development and Political Legitimacy." *American Political Science Review* 53(1):69–105.

Little, Andrew T. and Thomas Zeitzoff. 2017. "A Bargaining Theory of Conflict with Evolutionary Preferences." *International Organization* 71(3):523–557.

Long, Andrew G., Timothy Nordstrom and Kyeonghi Baek. 2007. "Allying for Peace: Treaty Obligations and Conflict Between Allies." *Journal of Politics* 69(4):1103–1117.

Luard, Evan. 1998. *A History of the United Nations. Vol. 1: The Years of Western Domination, 1945–1955*. London: Macmillan.

Luce, Edward. 2017. *The Retreat of Western Liberalism*. New York: Atlantic Monthly Press.

Lynch, Allen. 1987. *The Soviet Study of International Relations*. New York: Cambridge University Press.

Mansbach, Richard W. and John A. Vasquez. 1981. *In Search of Theory: A New Paradigm for Global Politics*. New York: Columbia University Press.

Mansfield, Edward D. and Brian M. Pollins. 2001. "The Study of Interdependence and Conflict: Recent Advances, Open Questions, and Directions for Future Research." *Journal of Conflict Resolution* 45(6): 834–859.

Mansfield, Edward D. and Brian M. Pollins, eds. 2006. *Economic Interdependence and International Conflict: New Perspectives on an Enduring Debate*. Ann Arbor: University of Michigan Press.

Maoz, Zeev and Bruce Russett. 1993. "Normative and Structural Causes of Democratic Peace, 1946–1986." *American Political Science Review* 87(3):624–638.

Marlantes, Karl. 2011. *What It Is Like to Go to War*. New York: Atlantic Monthly Press.

Massie, Robert K. 1991. *Dreadnought: Britain, Germany, and the Coming of the Great War*. New York: Random House.

Mattes, Michaela and Burcu Savun. 2009. "Fostering Peace After Civil War: Commitment Problems and Agreement Design." *International Studies Quarterly* 53(3):737–759.

Mattes, Michaela and Burcu Savun. 2010. "Information, Agreement Design, and the Durability of Civil War Settlements." *American Journal of Political Science* 54(2):511–524.

Matthews, Dylan. 2018. "35 Years Ago Today, One Man Saved Us From World-Ending Nuclear War." **URL:** *https://www.vox.com/2018/9/26/17905796/nuclear-war-1983-stanislav-petrov-soviet-union*

Mattis, James. 2018. "Summary of the 2018 National Defense Strategy of the United States of America: Sharpening the American Military's Competitive Edge." **URL:** *https://www.defense.gov/Portals/1/Documents/pubs/2018-National-Defense-Strategy-Summary.pdf*

McDonald, H. Brooke and Richard Rosecrance. 1985. "Alliance and Structural Balance in the International System: A Reinterpretation." *Journal of Conflict Resolution* 29(1):57–82.

McDonald, Patrick J. 2015. "Great Powers, Hierarchy, and Endogenous Regimes: Rethinking the Domestic Causes of Peace." *International Organization* 69(3):557–588.

McLaughlin Mitchell, Sara and Paul R. Hensel. 2007. "International Institutions and Compliance With Agreements." *American Journal of Political Science* 51(4):721–737.

McLaughlin Mitchell, Sara and Clayton L. Thyne. 2010. "Contentious Issues as Opportunities for Diversionary Behavior." *Conflict Management and Peace Science* 27(5):461–485.

McMeekin, Sean. 2013. *July 1914: Countdown to War*. New York: Basic Books.

Mearsheimer, John J. 1990. "Back to the Future: Instability in Europe After the Cold War." *International Security* 15(1):5–56.

Mearsheimer, John J. 2001. *The Tragedy of Great Power Politics*. New York: W.W. Norton.

Mearsheimer, John J. and Stephen M. Walt. 2013. "Leaving Theory Behind: Why Simplistic Hypothesis Testing Is Bad for International Relations." *European Journal of International Relations* 19(3):427–457.

Medlicott, William Norton. 1969. *Bismarck, Gladstone, and the Concert of Europe*. New York: Greenwood Press.

Milgram, Stanley. 1963. "Behavioral Study of Obedience." *Journal of Abnormal and Social Psychology* 67(4):371–378.

Milton, Giles. 2000. *Nathaniel's Nutmeg: Or, The True and Incredible Adventures of the Spice Trader Who Changed the Course of History*. New York: Penguin.

Mishra, Pankaj. 2017. *Age of Anger: A History of the Present*. New York: Farrar, Straus and Giroux.

Mitzen, Jennifer. 2013*a*. "The Irony of Pinkerism." *Perspectives on Politics* 11(2):525–528.

Mitzen, Jennifer. 2013*b*. *Power in Concert: The Nineteenth-Century Origins of Global Governance*. Chicago: University of Chicago Press.

Morgenthau, Hans J. 1948. *Politics Among Nations: The Struggle for Power and Peace*. New York: Alfred A. Knopf.

Morgenthau, Hans J. 1960. *Politics Among Nations: The Struggle for Power and Peace*. 3rd ed. New York: Alfred A. Knopf.

Morley, Morris H. 2002. *Washington, Somoza, and the Sandinistas: State and Regime in U.S. Policy Toward Nicaragua, 1969–1981*. New York: Cambridge University Press.

Morrow, James D. 1993. "Arms Versus Allies: Trade-Offs in the Search for Security." *International Organization* 47(2):207–233.

Mueller, Dennis C. 1989. *Public Choice II*. Cambridge: Cambridge University Press.

Mueller, John. 1989. *Retreat from Doomsday: The Obsolescence of Major War*. New York: Basic Books.

Mueller, John. 2004. "What Was the Cold War About? Evidence from Its Ending." *Political Science Quarterly* 119(4):609–631.

Mueller, John. 2007. *The Remnants of War*. Ithaca, NY: Cornell University Press.

Mueller, John. 2014. "Did History End? Assessing the Fukuyama Thesis." *Political Science Quarterly* 129(1):35–54.

Mueller, John E. 1970. "Presidential Popularity from Truman to Johnson." *American Political Science Review* 64(1):18–34.

Müller, Jan-Werner. 2016. *What Is Populism?* Philadelphia: University of Pennsylvania Press.

Murray, C. J. L., G. King, A. D. Lopez, N. Tomijima and E. G. Krug. 2002. "Armed Conflict as a Public Health Problem." *BMJ* 324(7333):346–349.

Nation, R. Craig. 1992. *Black Earth, Red Star: A History of Soviet Security Policy, 1917–1991*. Ithaca, NY: Cornell University Press.

Nexon, Daniel H. 2009. *The Struggle for Power in Early Modern Europe: Religious Conflict, Dynastic Empires, and International Change*. Princeton, NJ: Princeton University Press.

Nicolson, Harold. 1946. *The Congress of Vienna: A Study in Allied Unity, 1812–1822*. New York: Harcourt, Brace and Co.

Oka, Rahul C., Marc Kissel, Mark Golitko, Susan Guise Sheridan, Nam C. Kim and Agustín Fuentes. 2017. "Population Is the Main Driver of War Group Size and Conflict Casualties." *Proceedings of the National Academy of Sciences* 114(52):E11101–E11110.

Oneal, John R. and Bruce M. Russet. 1997. "The Classical Liberals Were Right: Democracy, Interdependence, and Conflict, 1950–1985." *International Studies Quarterly* 41(2):267–294.

Orend, Brian. 2000. *War and International Justice: A Kantian Perspective*. Waterloo, ON: Wilfrid Laurier University Press.

Organski, A. F. K. and Jacek Kugler. 1980. *The War Ledger*. Chicago: University of Chicago Press.

Osborne, Roger. 2006. *Civilization: A New History of the Western World*. New York: Pegasus.

Owsiak, Andrew P. 2012. "Signing Up for Peace: International Boundary Agreements, Democracy, and Militarized Interstate Conflict." *International Studies Quarterly* 56(1):51–66.

Oye, Kenneth A., ed. 1986. *Cooperation Under Anarchy*. Princeton, NJ: Princeton University Press.

Palmer, G., V. D'Orazio, M. Kenwick, and M. Lane. 2015. "The MID4 Dataset, 2002–2010: Procedures, Coding Rules, and Description." *Conflict Management and Peace Science* 32(2):222–242.

Peterson, T. M. 2011. "Third-Party Trade, Political Similarity, and Dyadic Conflict." *Journal of Peace Research* 48(2):185–200.

Peterson, Timothy M. and Cameron G. Thies. 2012. "Beyond Ricardo: The Link Between Intra-Industry Trade and Peace." *British Journal of Political Science* 42(04):747–767.

Peterson, Timothy M. and Stephen L. Quackenbush. 2010. "Not All Peace Years Are Created Equal: Trade, Imposed Settlements, and Recurrent Conflict." *International Interactions* 36(4):363–383.

Pettersson, Therése and Kristine Eck. 2018. "Organized Violence, 1989–2017." *Journal of Peace Research* 55(4):535–547.

Pevehouse, Jon and Bruce Russett. 2006. "Democratic International Governmental Organizations Promote Peace." *International Organization* 60(4):969–1000.

Phillips, Andrew and J. C. Sharman. 2015. *International Order in Diversity: War, Trade and Rule in the Indian Ocean*. Cambridge: Cambridge University Press.

Pillar, Paul R. 1983. *Negotiating Peace: War Termination as a Bargaining Process*. Princeton, NJ: Princeton University Press.

Pinheiro, João, Eugénia Cunha and Steven Symes. 2015. Over-Interpretation of Bone Injuries and Implications for Cause and Manner of Death. In *Skeletal Trauma Analysis*, edited by Nicholas V. Passalacqua and Christopher W. Rainwater, 27–41. Chichester: John Wiley.

Pinker, Steven. 2011*a*. *The Better Angels of Our Nature: Why Violence Has Declined*. New York: Viking Adult.

Pinker, Steven. 2011*b*. "Violence Vanquished." *Wall Street Journal*, September 24.
URL: *https://www.wsj.com/articles/SB10001424053111904106704576583203589408180?mod=WSJ_hp_LEFTTopStories*

Pinker, Steven. 2012. "Fooled by Belligerence: Comments on Nassim Taleb's 'The Long Peace Is a Statistical Illusion.'" Online manuscript.
URL: *http://stevenpinker.com/files/comments_on_taleb_by_s_pinker.pdf*

Pinker, Steven. 2015*a*. "Now for the Good News: Things Really Are Getting Better." *The Guardian*, September 11.
URL: *https://www.theguardian.com/commentisfree/2015/sep/11/news-isis-syria-headlines-violence-steven-pinker*

Pinker, Steven. 2015*b*. "Response to the Book Review Symposium: Steven Pinker, *The Better Angels of Our Nature*." *Sociology* 49(4):NP3–NP8.

Pinker, Steven. 2017. "Has the Decline of Violence Reversed Since *The Better Angels of Our Nature* Was Written?"
URL: *https://stevenpinker.com/files/pinker/files/has_the_decline_of_violence_reversed_since_the_better_angels_of_our_nature_was_written_2017.pdf*

Pinker, Steven. 2018. *Enlightenment Now: The Case for Reason, Science, Humanism, and Progress*. New York: Viking.

Powell, Robert. 2006. "War as a Commitment Problem." *International Organization* 60(1):169–203.

Putnam, Robert D. 1995. "Bowling Alone: America's Declining Social Capital." *Journal of Democracy* 6(1):65–78.

Quackenbush, S. L. and J. F. Venteicher. 2008. "Settlements, Outcomes, and the Recurrence of Conflict." *Journal of Peace Research* 45(6):723–742.

Quackenbush, Stephen. 2006. "Identifying Opportunity for Conflict: Politically Active Dyads." *Conflict Management and Peace Science* 23(1):37–51.

Raleigh, Clionadh and Dominic Kniveton. 2012. "Come Rain or Shine: An Analysis of Conflict and Climate Variability in East Africa." *Journal of Peace Research* 49(1):51–64.

Raska, Michael. 2014. "What Is China's Strategy for North Korea?"
URL: *http://www.aljazeera.com/indepth/opinion/2014/05/what-china-strategy-north-korea-201451473945985671.html*

Rath, Arun. 2014. "The Summer of 2014 Has Been a Messy Time for the World."
URL: *https://www.npr.org/2014/09/13/348286463/the-summer-of-2014-has-been-a-messy-time-for-the-world*

Reed, William, David H. Clark, Timothy Nordstrom and Wonjae Hwang. 2008. "War, Power, and Bargaining." *Journal of Politics* 70(4):1203–1216.

Reiter, Dan, Allan C. Stam and Michael C. Horowitz. 2016. "A Revised Look at Interstate Wars, 1816–2007." *Journal of Conflict Resolution* 60(5):956–976.

Reus-Smit, Christian. 2017. "Cultural Diversity and International Order." *International Organization* 71(4):851–885.

Richards, Diana, ed. 2000. *Political Complexity: Nonlinear Models of Politics*. Ann Arbor: University of Michigan Press.

Richardson, Lewis F. 1960*a*. *Arms and Insecurity*. Pittsburgh: Boxwood Press and Quadrangle.

Richardson, Lewis F. 1960*b*. *Statistics of Deadly Quarrels*. Chicago: Quadrangle Books.

Rider, Toby J. 2009. "Understanding Arms Race Onset: Rivalry, Threat, and Territorial Competition." *Journal of Politics* 71(2):693–703.

Roberts, Priscilla Mary. 2012. Arkhipov, Vasili Alexandrovich. In *Cuban Missile Crisis: The Essential Reference Guide*, 13–14. Santa Barbara, CA: ABC-CLIO.

Rosenau, James N. 1966. "Pre-Theories and Theories of Foreign Policy." In *Approaches to Comparative and International Politics*, ed. R. Barry Farrell. Evanston, IL: Northwestern University Press.

Ross, Lee. 1977. "The Intuitive Psychologist and His Shortcomings: Distortions in the Attribution Process." In *Advances in Experimental Social Psychology. Volume 10*, edited by Leonard Berkowitz, 173–220. New York: Academic Press.

Ross, Lee and Richard E. Nisbett. 1991. *The Person and the Situation: Perspectives of Social Psychology*. Philadelphia, PA: Temple University Press.

Rotberg, Robert I. 2004. *When States Fail: Causes and Consequences*. Princeton, NJ: Princeton University Press.

Rousseau, David, Christopher Gelpi, Dan Reiter and Paul Huth. 1996. "Assessing the Dyadic Nature of the Democratic Peace, 1918–88." *American Political Science Review* 90(3):512–533.

Russett, Bruce. 1993. *Grasping the Democratic Peace: Principles for a Post–Cold War World*. Princeton, NJ: Princeton University Press.

Russett, Bruce and John R. Oneal. 2001. *Triangulating Peace: Democracy, Interdependence, and International Organizations*. New York: Norton.

Russett, Bruce, John Oneal and David R. Davis. 1998. "The Third Leg of the Kantian Tripod for Peace: International Organizations and Militarized Disputes, 1950–1985." *International Organization* 52(3):441–467.

Ryan, David. 1995. *US-Sandinista Diplomatic Relations: Voice of Intolerance*. New York: St. Martin's Press.

Salehyan, Idean. 2008. "From Climate Change to Conflict? No Consensus Yet." *Journal of Peace Research* 45(3):315.

Salehyan, Idean and Cullen S. Hendrix. 2014. "Climate Shocks and Political Violence." *Global Environmental Change* 28:239–250.

Sambanis, Nicholas and Moses Shayo. 2013. "Social Identification and Ethnic Conflict." *American Political Science Review* 107(2):294–325.

Sample, Susan G. 1997. "Arms Races and Dispute Escalation: Resolving the Debate." *Journal of Peace Research* 34(1):7–22.

Santayana, George. 1922. *Soliloquies in England and Later Soliloquies*. New York: Charles Scribner's Sons.

Sarkees, Meredith Reid. n.d. "The COW Typology of War: Defining and Categorizing Wars (Version 4 of the Data)." **URL:** *http://tinyurl.com/nzzu5dj*

Schroeder, Paul W. 1994. *The Transformation of European Politics*. Oxford: Clarendon Press.

Schultz, Kenneth. 1999. "Do Democratic Institutions Constrain or Inform? Contrasting Two Institutional Perspectives on Democracy and War." *International Organization* 53(2):233–266.

Schultz, Kenneth A. 2014. "What's in a Claim? De Jure Versus De Facto Borders in Interstate Territorial Disputes." *Journal of Conflict Resolution* 58(6):1059–1084.

Schweller, Randall L. 1998. *Deadly Imbalances: Tripolarity and Hitler's Strategy of World Conquest*. New York: Columbia University Press.

Schweller, Randall L. 2014. *Maxwell's Demon and the Golden Apple: Global Discord in the New Millennium*. Baltimore, MD: Johns Hopkins University Press.

Schweller, Randall L. 2016. "The Balance of Power in World Politics." In *Oxford Research Encyclopedia of Politics*, edited by William R. Thompson. Vol. 1. New York: Oxford University Press.

Scott, George. 1973. *The Rise and Fall of the League of Nations*. London: Hutchinson.

Shannon, Megan. 2009. "Preventing War and Providing the Peace? International Organizations and the Management of Territorial Disputes." *Conflict Management and Peace Science* 26(2):144–163.

Shannon, Megan, Daniel Morey and Frederick J. Boehmke. 2010. "The Influence of International Organizations on Militarized Dispute Initiation and Duration." *International Studies Quarterly* 54(4):1123–1141.

Simmons, Lee. 2013. "Want to Save Lives? You Need a Map of What's Doing Us In." *Wired*. **URL:** *https://www.wired.com/2013/11/infoporn-causes-of-death/*

Singer, Peter. 2011. "Is Violence History?" *New York Times* October 9. **URL:** *http://www.nytimes.com/2011/10/09/books/review/the-better-angels-of-our-nature-by-steven-pinker-book-review.html*

Slantchev, Branislav L. 2005. "Territory and Commitment: The Concert of Europe as Self-Enforcing Equilibrium." *Security Studies* 14(4):565–606.

Slater, Joanna. 2017. "Goodbye to All That: Is the International Order as We Know It Over?" *Globe and Mail*. **URL:** *https://www.theglobeandmail.com/news/world/goodbye-to-all-that-is-the-international-order-as-we-know-itover/article34826488/*

Slaughter, Anne Marie. 2004. *A New World Order*. Princeton, NJ: Princeton University Press.

Small, Melvin and J. David Singer. 1982. *Resort to Arms: International and Civil Wars, 1816–1980*. Beverly Hills, CA: SAGE.

Smith, Adam and Edwin Cannan. 1994. *An Inquiry Into the Nature and Causes of the Wealth of Nations*. New York: Modern Library.

Smith, David. 2011. "Ivory Coast: UN's Intervention Broke the Impasse." *The Guardian*. **URL:** *https://www.theguardian.com/world/2011/apr/05/ivory-coast-un-intervention*

Snyder, Glenn H. 1984. "The Security Dilemma in Alliance Politics." *World Politics* 36(4):461–495.

Snyder, Jack. 1993. "The New Nationalism: Realist Interpretations and Beyond." In *The Domestic Bases of Grand Strategy*, edited by Richard Rosecrance and Arthur A. Stein, 179–200. Ithaca, NY: Cornell University Press.

Snyder, Jack L. and Robert Jervis, eds. 1993. *Coping with Complexity in the International System*. Boulder, CO: Westview Press.

Solingen, Etel. 1998. *Regional Orders at Century's Dawn: Global and Domestic Influences on Grand Strategy*. Princeton, NJ: Princeton University Press.

Sontag, Raymond. 1938. *Germany and England: Background of Conflict, 1848–1894*. New York: D. Appelton - Century.

Spagat, Michael and Steven Pinker. 2016. "Warfare." *Significance* 13(3):44. **URL:** *http://onlinelibrary.wiley.com/doi/10.1111/j.1740-9713.2016.00903.x/full*

Spagat, Michael and Stijn van Weezel. 2018. "On the Decline of War." In *UDC Centre for Economic Research Working Paper Series*. Dublin. **URL:** *http://www.ucd.ie/t4cms/WP18_15.pdf*

Spiro, Peter. 2014. "Ukraine, International Law, and the Perfect Compliance Fallacy." **URL:** *http://opiniojuris.org/2014/03/02/ukraine-international-law-perfect-compliance-fallacy/*

Spruyt, Hendrik. 1994. *The Sovereign State and Its Competitors: An Analysis of Systems Change*. Princeton, NJ: Princeton University Press.

Steinberg, Jonathan. 2011. *Bismarck: A Life*. New York: Oxford University Press.

Stewart-Ingersoll, Robert and Derrick Frazier. 2012. *Regional Powers and Security Orders: A Theoretical Framework*. New York: Routledge.

Stoll, Clifford. 1989. *The Cuckoo's Egg: Tracking a Spy Through the Maze of Computer Espionage*. New York: Doubleday.

Suro, Roberto. 1987. "On a Gulf Tanker: Waiting for the Worst." *New York Times* November 8.

Suzuki, Shogo, Yongjin Zhang and Joel Quirk, eds. 2016. *International Orders in the Early Modern World: Before the Rise of the West*. New York: Routledge.

Taleb, Nassim Nicholas. 2007. *The Black Swan: The Impact of the Highly Improbable*. New York: Random House.

Taleb, Nassim Nicholas. 2012. "The 'Long Peace' Is a Statistical Illusion." Online manuscript. **URL:** *https://web.archive.org/web/20121117225617/http://www.fooledbyrandomness.com/longpeace.pdf*

Taleb, Nassim Nicholas. 2015. "Additional Comments on Cirillo and Taleb (2014)." **URL:** *http://www.fooledbyrandomness.com/comment.pdf*

Taylor, A. J. P. 1954. *The Struggle for Mastery in Europe, 1848–1918*. New York: Oxford University Press.

Taylor, A. J. P. 1967. *Bismarck: The Man and Statesman*. New York: Vintage.

Thomson, David. 1978. *Europe Since Napoleon*. New York: Penguin Books.

Thucydides. 1972 (431 B.C.E.). *History of the Peloponnesian War*. 2nd ed. London: Penguin.

Tilly, Charles. 1975. "Reflections on the History of European State-Making." In *The Formation of National States in Western Europe*, edited by Charles Tilly, 3–83. Princeton, NJ: Princeton University Press.

Tilly, Charles. 1992. *Coercion, Capital, and European States, AD 990–1992*. Cambridge, MA: Blackwell.

Tingley, Dustin H. 2011. "The Dark Side of the Future: An Experimental Test of Commitment Problems in Bargaining." *International Studies Quarterly* 55(2):521–544.

Tir, Jaroslav. 2005. "Keeping the Peace After Secession: Territorial Conflicts Between Rump and Secessionist States." *Journal of Conflict Resolution* 49(5):713–741.

Tir, Jaroslav and Douglas M. Stinnett. 2012. "Weathering Climate Change: Can Institutions Mitigate International Water Conflict?" *Journal of Peace Research* 49(1):211–225.

Toland, John. 2003. *The Rising Sun: The Decline and Fall of the Japanese Empire, 1936–1945*. New York: Modern Library.

Tomz, Michael R. and Jessica L. P. Weeks. 2013. "Public Opinion and the Democratic Peace." *American Political Science Review* 107(4):849–865.

Turchin, Peter. 2007. *War and Peace and War: The Rise and Fall of Empires*. New York: Plume.

Turchin, Peter. 2016. *Ultrasociety: How 10,000 Years of War Made Humans the Greatest Cooperators on Earth*. Chaplin, CT: Beresta Books.

Ulam, Adam B. 1974. *Expansion and Coexistence: Soviet Foreign Policy, 1917–73*. New York: Praeger.

United Nations High Commissioner for Refugees. 2017. "UNHCR Global Trends 2017." **URL:** *https://www.unhcr.org/statistics/unhcrstats/5b27be547/unhcr-global-trends-2017.html*

United Nations Operations in Côte d'Ivoire. n.d. "Post-Election Crisis." **URL:** *http://www.un.org/en/peacekeeping/missions/unoci/elections.shtml*

Uslaner, Eric M. 1976. "The Pitfalls of Per Capita." *American Journal of Political Science* 20(1):125–133.

Van Evera, Stephen. 1985. The Cult of the Offensive and the Origins of the First World War. In *Military Strategy and the Origins of the First World War*, edited by Steven E. Miller, 58–107. Princeton, NJ: Princeton University Press.

Van Evera, Stephen. 1990. "Primed for Peace: Europe After the Cold War." *International Security* 15(3):7–57.

Van Evera, Stephen. 1994. "Hypotheses on Nationalism and War." *International Security* 18(4):5–39.

Vasquez, John A. 1995. "Why Do Neighbors Fight? Proximity, Interaction, or Territoriality." *Journal of Peace Research* 32(3):277–293. **URL:** *http://jpr.sagepub.com/content/32/3/277.short*

Väyrynen, Raimo. 2006. *The Waning of Major War: Theories and Debates*. New York: Routledge.

Voeten, Erik. 2005. "The Political Origins of the UN Security Council's Ability to Legitimize the Use of Force." *International Organization* 59(3):527–557.

Von Bortkewitsch, Ladislas. 1898. *Das Gesetz Der Kleinen Zahlen*. Leipzig: B. G. Teubner.

Von Clausewitz, Carl. 1976 (1832). *On War*. Princeton, NJ: Princeton University Press.

Wagner, R. Harrison. 2000. "Bargaining and War." *American Journal of Political Science* 44(3):469–484.

Wallace, Michael D. 1982. "Armaments and Escalation: Two Competing Hypotheses." *International Studies Quarterly* 26(1):37–56.

Waltz, Kenneth N. 1959. *Man, the State, and War: A Theoretical Analysis.* New York: Columbia University Press.

Waltz, Kenneth N. 1964. "The Stability of a Bipolar World." *Daedalus* 93(3):881–909.

Waltz, Kenneth N. 1967. *Foreign Policy and Democratic Politics.* Boston: Little, Brown and Co.

Waltz, Kenneth N. 1979. *Theory of International Politics.* New York: Random House.

Warren, T. Camber. 2016. "Modeling the Coevolution of International and Domestic Institutions: Alliances, Democracy, and the Complex Path to Peace." *Journal of Peace Research* 53(3):424–441.

Wawro, Gregory J. and Ira Katznelson. 2014. "Designing Historical Social Scientific Inquiry: How Parameter Heterogeneity Can Bridge the Methodological Divide Between Quantitative and Qualitative Approaches." *American Journal of Political Science* 58(2):526–546.

Weisiger, Alex. 2013. *Logics of War: Explanations for Limited and Unlimited Conflicts.* Ithaca, NY: Cornell University Press.

Wendt, Alexander. 1992. "Anarchy Is What States Make of It: The Social Construction of Power Politics." *International Organization* 46(2):395–421.

Werner, S. 1998. "Negotiating the Terms of Settlement: War Aims and Bargaining Leverage." *Journal of Conflict Resolution* 42(3):321–343.

Westra, Joel H. 2010. "Cumulative Legitimation, Prudential Restraint, and the Maintenance of International Order: A Re-Examination of the UN Charter System." *International Studies Quarterly* 54(2):513–533.

White, Matthew. 2013. *Atrocities: The 100 Deadliest Episodes in Human History.* New York: W. W. Norton.

Wilkenfeld, Jonathan and Michael Brecher. 1989. "International Crises, 1945–1975: The UN Dimension." In *The Politics of International Organizations: Patterns and Insights*, edited by Paul F. Diehl, 154–173. Chicago: Dorsey Press.

Williams, Howard. 2012. *Kant and the End of War: A Critique of Just War Theory.* New York: Palgrave Macmillan.

Williams, John T. and Michael D. McGinnis. 1988. "Sophisticated Reaction in the U.S.-Soviet Arms Race: Evidence of Rational Expectations." *American Journal of Political Science* 32(4):968–995.

Wilson, William J. 1978. *The Declining Significance of Race: Blacks and Changing American Institutions.* Chicago: University of Chicago Press.

Wilson, William Julius. 2011. "*The Declining Significance of Race*: Revisited & Revised." *Daedalus* 140(2):55–69.

Wittkopf, Eugene R. 1990. *Faces of Internationalism: Public Opinion and American Foreign Policy.* Durham, NC: Duke University Press.

Wohlforth, William Curti. 1993. *The Elusive Balance: Power and Perception During the Cold War.* Ithaca, NY: Cornell University Press.

Wolford, Scott. 2007. "The Turnover Trap: New Leaders, Reputation, and International Conflict." *American Journal of Political Science* 51(4):772–788. **URL:** *http://onlinelibrary.wiley.com/doi/10.1111/j.1540-5907.2007.00280.x/full*

Wright, Thorin M. and Toby J. Rider. 2014. "Disputed Territory, Defensive Alliances, and Conflict Initiation." *Conflict Management and Peace Science* 31(2):119–144.

Zarakol, Ayşe, ed. 2017. *Hierarchies in World Politics.* New York: Cambridge University Press.

Zimmerman, William. 1969. *Soviet Perspectives on International Relations, 1956–1967.* Princeton, NJ: Princeton University Press.

Zimmerman, William. 1973. "Issue Area and Foreign-Policy Process: A Research Note in Search of a General Theory." *American Political Science Review* 67(4):1204–1212.

Zubok, Vladislav and Constantine Pleshakov. 1996. *Inside the Kremlin's Cold War.* Cambridge, MA: Harvard University Press.

INDEX

For the benefit of digital users, indexed terms that span two pages (e.g., 52–53) may, on occasion, appear on only one of those pages.

Note: page numbers followed by *f* and *t* refer to figures and tables respectively. Those followed by n refer to notes, with note number.